Basic Concept of Physical Chemistry

Basic Concept of Physical Chemistry

P.K. Mohanti

CYBER TECH PUBLICATIONS
4264/3, Ansari Road, Daryaganj, New Delhi-110002 (India)
Ph.: 011-23244078. Fax: 011-23280028
E-Mail: cyberpublicationsdelhi@yahoo.com

BASIC CONCEPT OF PHYSICAL CHEMISTRY

P.K. Mohanti

First Edition 2009

Published by :

G.S. Rawat for **Cyber Tech Publications**
4264/3, Ansari Road, Daryaganj, New Delhi-110002 (India)
Ph.: 011-23244078. Fax: 011-23280028
E-Mail: cyberpublicationsdelhi@yahoo.com

Printed at : Asian Offset Printers

PREFACE

This book '*Basic Concept of Physical Chemistry*' to cater to the need of graduate and post-graduate students. Since I do not consider the chemistry of natural products to be fundamental chemistry but rather the application of fundamental principles, I have largely excluded the study of these compounds.

Even so, it is my belief that in order to learn organic chemistry it is necessary to acquire a comprehensive knowledge of chemical behaviour and then, with this as a background, to consider the principles and the theoretical basis of the subject. To many beginners organic chemistry may seem to consist of a large variety of methods and reactions which appear isolated and, consequently, only to be learned by heart. However, the introduction of the electronic theories of organic chemistry as early as possible and a constant application of their principles give to organic chemistry a certain coherence that is soon appreciated by the beginner, and thus facilitate his study in this branch of chemistry.

Contents

1

An Introduction to Physical Chemistry

Whenever a number of phenomenon have been observed and studied with exact measurements, we develop a law that will predict the behaviour of similar systems under different conditions. A concise statement which summerizes a large number of related facts is called law, *e.g.*, law of definite proportions, law of multiple proportions, Boyles law, etc. Natural laws may be discovered either by correlation of experimentally determined facts or by means of a guess based on a few facts which can be used as the basis of probable cause of phenomenon or for planning experiments to collect more information. The science of transformation of substances is known as chemistry.

Nature never repeats itself. In order to copy nature, certain phenomenon are artificially reproduced. These are called experiments. They may be the reproduction of natural of artificial of natural or artificial phenomenon. The differentiation between science and philosophy started after 1690, when John Locke termed science as the study of corpuscular mechanism while philosophy explored the mind and its ideas. Nowadays the physical science are taking physical world as their domain and philosophy is exploring the mental world of thoughts and ideas.

Such a guess regarding the cause of the phenomenon, is called hypothesis, *e.g.*, Avogadro's hypothesis has been subjected to the test of experiments and has been shown to have very wide applications, it is termed as theory. A theory is more formal and based on more facts than hypothesis. Many hypotheses are destined to be discarded when new facts are revealed but they play a very vital role in the development of science.

A successful hypothesis is one that stimulates additional research, opens up new fields or explains and coordinates previously unrelated facts.

CONCENTRATION

Concentration is expressed in several forms provided there is no chemical reaction between the solute and solvent.

Molar

Number of gram molecular weight (mole) of the solute dissolved to give 1 litre of the solution is the molarity of the solution and is designated by M. If C_i is the number of mole n_i of i species per unit volume of the solution of total volume V_m and P_i is the number density of the particle of species i, then

$$Ci = \frac{n_i}{V_m} = \frac{P_i}{A} \qquad ...(1)$$

Molal

The number of moles of the solute dissolved in 1000 g of the solvent is known as molal solution and is designated by m.

$$m_i = \frac{n_i}{W_A} = \frac{W_i \times 1000}{M_i W_A} \qquad ...(2)$$

where W_A is the weight of solvent in g.

Mole Fraction

The mole fraction is the ratio of the number of moles of one component to the total number of moles in the solution. If we dissolve 1 mole of sugar in 4 moles of water, the mole fraction of sugar will be

$$\frac{1x_i}{5} = \frac{n_i}{n} \cdot \text{where } x_i \text{ is the mole fraction.} \qquad ...(3)$$

$$x_i = \frac{n_i}{\sum_{J=1}^{c} n_j} \text{ where c is the total components.} \qquad ...(4)$$

Normal Solution

The normality (N) of a solution is the number of gram equivalent weights of solute per litre of the solution.

Gram equivalent weight = *mole weight* (No. of available H^{+1})

or OH^{-1} per molecule ...(5)

Thus, one mole of HCl = 0.5 mole of H_2SO_4 = 0.33 mole of H_3PO_4

1 mole of NaOH = 0.5 mole of $Ca(OH)_2$

Formal Solution

The number of gram formulae of the solute per litre of the solution is known as the formality of the solution and is designated as *F.*

$$F_i = \frac{W_i}{M_i} \times \frac{1000}{V} \qquad ...(6)$$

The terms generally used for the expression of concentration are either normality or molarity.

PROCESS

Any change in the system is called a process.

Irreversible Process

All real, actual, spontaneous and naturally occurring change, tending to proceed in a certain direction and when left to themselves do not go in opposite direction, are called irreversible processes, *e.g.*, march of time, water falling in a water fall, a falling weight or a chemical reaction which changes in chemical or physical properties.

These are always spontaneous processes. The end point of any process is said to be in a state of *equilibrium.* Any system in equilibrium is said to be in a stable state.

Reversible Process

If a process is carried out in such a way that the system throughout the process in a state of equilibrium, *i.e.*, consists of a series of successive equilibria, the process is known as *reversible process.*

A reversible process is one which is performed in such a way that at any moment, by changing an independent variable by an infinitesimal amount, the process is reversed, *e.g.*, to expand a gas reversibly, the pressure on the piston is removed infinitely slowly so that at every instant the pressure everywhere in the gas volume is exactly the same and is just equal to the opposing pressure of the piston. Such are only ideal processes which can be approach but not actually reached in practice.

ENERGY

To bring about the change in system it is essential either the energy should be supplied or taken out. Therefore, the *energy* of a system is that body's or system's capacity to work. This is given by *force* applied on a system in moving is by distance. Issac Newton showed in 1642 that the *force* is given by the product of the *mass* of the body and the *acceleration*. Let a force f_x act on a particle of mass m, moving with a velocity of v is the x direction at the time t.

$$f_x = \frac{md^2x}{dt^2} = m\frac{dv}{dt} = mv\frac{dv}{dt} \qquad ...(1)$$

where $\frac{d^2x}{dv^2}$ or $\frac{dv}{dx}$ is called as acceleration and mv as momentum.

On integration of equation 1

$$\frac{1}{2}mv^2 \int fdx = E$$

where E is the total energy.

On the left the first term is the kinetic energy and the second term is the potential energy u, *i.e.*, the work which must be done on the particle in moving it counter to the force.

$$U = -\int f_x dx \text{ or } f_x = -\frac{dU}{dx}$$

Mass into velocity is called *momentum*. The change in momentum is force with the mass in grams, time in seconds and the displacement in centimetres (CGS system). The unit of force is dyne. With mass in kilograms, time in seconds and the displacement in metres, the unit of force is newton. Newtons law of gravitation states that there is an attractive force between two masses m_1 and m_2 proportional to their product and inversely proportional to the square of the distance between them.

$$f = k\frac{m_1m_2}{d_{12}^2}$$

where k is the proportionality constant and has a value 6.66×10^{-8} cm^3 sec^{-2} g^{-1} and d_{12} is the distance between m_1 and m_2.

Internal Energy

The energy stored in the system is a definite property of the system and is called energy content or internal energy of the system and is denoted by E. It includes all the energies of translational, rotational, and vibrational motion, surface energies and energies due to the attractive and repulsive forces acting between atoms, molecules and sub atomic particles, but it does not include energies due to the position under the influence of several types of fields or kinetic energy due to mass motion of the system. It is an extensive property and as such independent of the path followed by the system. It depends only on the quantity of matter. It is not possible to determine the internal energy of a system but a change in internal energy can always be evaluated by the heat absorbed (h) and the work (w). As a convention heat absorbed by the system from the surroundings is always positive h (melting of ice). w is also positive when the work is done by the system on the surroundings. The work may be mechanical, electrical, or on surface by changing the area. In an ideal gas a work at constant pressure is

$$\Delta E = h - w \qquad ...(2)$$

This is known as the first law of thermodynamics

$$\Delta E = h - P \Delta V \qquad ...(3)$$

where P is the pressure and DV is the change in volume.

By rearranging

$$H = E + PV \qquad ...(4)$$

H is known as the enthalpy of the system which is also a function of state.

Entropy

Similarly, in a reversible process if ∂h is the infinitesimal heat absorbed by the system from isothermal surroundings at a temperature T, then

$$\partial s = \partial h/T \text{ or } dH = T\Delta S \qquad ...(5)$$

s is known as entropy of the system. This is again an extensive property and is a function of state of the system.

Free Energy

In a constant pressure process G = H– Ts where G is known as Gibbs free energy. In a constant volume process A = E – Ts. Where A is known as Helmholtz free energy. In a solution the free energy is a function of temperature, pressure and the number of moles of each of the reagent present. Therefore, in an open system

$$\mu_i = \left(\frac{\partial G}{\partial n_i}\right)_{T,P.nj} \qquad ...(6)$$

$$= \left(\frac{\partial E}{\partial n_i}\right)_{V.S.nj} = G_i$$

where μ_i is known as the chemical potential of the *i*th component, keeping pressure, temperature and other components constant. This is the rate of change in Gibbs free energy G_i of the system with the number of moles n_i keeping other parameters constant.

Laws of Conservation of Energy and Mass

The energy is either stored in the system or passes freely from one system to another either in the same form or in other forms. Helmholtz in 1847, stated that energy is neither created not destroyed in any transformation of matter. This is known as the law of conservation of energy. In drilling a hole inn a metal, work is converted into heat. In a storage batter, chemical energy is converted into electrical energy.

Lomonosov in 1756 and Lavoisir in 1783, propounded that mass is neither created not destroyed in any transformation of matter. This is known as the law of conservation of mass. In burning a candle the loss in weight of candle is equivalent to the formation of carbon dioxide and water. In an atomic explosion, the loss in mass is equivalent to huge amount of energy generated.

According to the modern concept, mass and energy together are conserved. It is the sum of mass and energy that is constant in any transformation.

Classification of Energy

There are a number of varieties of energy. Potential energy, kinetic energy, heat energy, electrical energy, radiation energy, chemical energy, atomic energy, surface energy are some of the common types.

Potential Energy

It is the energy that a system possess by virtue of its position in a field of force and its particular configuration. Depending upon the type of the field of force, it is termed as gravitational energy, surface energy, heat energy, mechanical energy, electrical energy, chemical energy, atomic energy, radiant energy, etc. On integration of equation 1.14, we obtain its total energy

$$E = \frac{1}{2}mv^2 + \int f_x d_x \text{ or } f_x = -\frac{dU}{dx}$$

where U is the potential energy which gives the work done on the particle in moving it counter to the force.

Kinetic Energy

The energy of a system by virtue of its motion is called kinetic energy. A moving automobile, an aeroplane in flight, all possess kinetic energy. It is quantitatively defined as half the mass into the square of the velocity $\left(\frac{1}{2}mc^2\right)$. Therefore, a change in the velocity of a moving body has a greater effect on its kinetic energy than the proportional changes in its mass. Depending upon the type of motion, it is also termed as translational, rotational, vibrational and electronic energy.

System

For the complete description of the system, at least one capacity factor mass is required. In the case of phase equilibrium only the intensity factors, *e.g.*, temperature, pressure and concentration are considered. The number of intensive state variables that can be independently varied without changing the number of *degrees of freedom* of the system.

In water –vapour system only one variable will determine the state. This system is known to have one degree of freedom. A certain amount of pure gas requires only two variables to be fixed, say pressure and temperature. The third is automatically fixed.

This will have two degrees of freedom. The position of a molecule as a whole in space may be described by giving the three coordinates of its centre of mass, but it is sufficient to specify temperature for a liquid.

The term phase can be defined structurally and thermodynamically. Thermodynamically a phase is a part of a system separated from others parts by interfaces and differing from them in thermodynamic properties. It must have sufficient volume for the concept of pressure, temperature and other thermodynamic properties like internal energy, heat content, etc. Phases should be separable from one another. Structurally phases differ in the order of mutual arrangement of their molecules.

Depending on this order one distinguishes between gases, liquids, solids,-crystalline, amorphous, plastics and elastomers.

In case there are more than one phases, than the smallest number of substances, in terms of which the composition of each of the phases in the system may be described separately is called *components.* There may be several possible choices of components which are equally satisfactory. The choice of the components is arbitrary, but the number of components is an important characteristic of a system. If calcium oxide, carbon dioxide and calcium carbonate are in equilibrium, the number of independent components is reduced by one because of the equilibrium expression

$$CaCO_3 = CaO + CO_2$$

Translational Kinetic Energy

The average translational energy for one mole of gas is 3/2 RT and is independent of the molecule. If all the thermal energy of a gas in the form translational energy we can write:

$$E_k = 3/2\ RT \qquad ...(7)$$

where E_k is average kinetic energy.

The degrees of freedom in a molecule are given by the number of coordinates required to locate all the mass points (atoms) in a molecule. If a molecule contains only one atom, it has three degree of freedom corresponding to translational motion in three directions X, Y, Z. According to the law of equipartition of energy, each translational degree of freedom in a molecule contributes 1/2 RT.

$$Ek = 1/2\ NM\bar{2} = 1/2\ NM\ (U\ \bar{2} + V\ \bar{2} + W\bar{2}) \qquad ...(8)$$

where U, V and W are the velocities in X, Y, Z directions and N & M are Avogadro number and Molecular weight respectively. If a molecule contains n atoms, each atom contributes these three degrees of freedom, so the molecule has a total 3n degrees of freedom.

Out of these 3 coordinates will be required to represent the instantaneous position of the centre of the mass. The remaining (3n–3) coordinates represent the internal degrees of freedom. In the case of a diatomic and polyatomic molecules kinetic energy is also associated with rotational and vibrational motions. In the case of diatomic molecule

$$3n - 3 = 3.$$

Rotational Energy (Diatomic Molecules)

The diatomic molecule has two atoms of mass m_1, and m_2 separated by a distance r (Fig. 1.1). Let them rotate about the axis z or x. Then the molecule will have a moment of inertia I given by

$$I = m_1r_1^2 + m_2 r_2^2 \qquad ...(9)$$

such that $$r_1 = \frac{m_2}{m_1 + m_2} r \text{ and } r_2 = \frac{m_1}{m_1 + m_2} r \qquad ...(10)$$

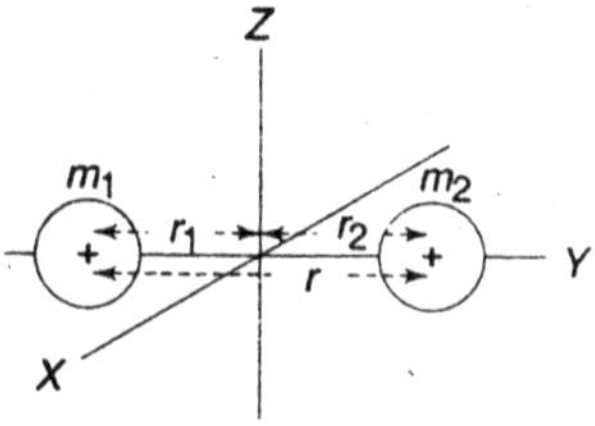

Fig. 1.1

Therefore, the moment of inertia I will be given by $\frac{m_1m_2}{m_1 + m_2} r_2 = \mu r^2$. The energy of rotation will be given by 1/2 $I\omega^2$ where ω is the angular velocity of rotation. μ is called the reduced mass of the molecule

$$m = \frac{m_1m_2}{m_1 + m_2} \qquad ...(11)$$

Only two coordinates are required to describe such a rotation completely, *e.g.*, two angles θ and ϕ (Fig. 1.1) will suffice to fix the orientation of the rotor in space. Thus, there are two degrees of freedom of rotation of a diatomic rotor, and average molar rotational energy will be given by RT (2N × 1/2 kT).

Vibrational Energy (Diatomic Molecule)

The vibrational motion in molecules is associated with both kinetic and potential energy, Fig. 1.6 shows a simple harmonic oscillator. The restoring force is directly proportional to its distance from the equilibrium position ($x = r - r_e$).

$$f = -kx = m\frac{d^2x}{dt^2} \qquad ...(12)$$

where k is the force constant.

The motion of a particle may be represented by potential energy function U(x)

$$f = -(\partial u/\partial x) = -kx$$

or $$U(x) = \frac{1}{2} kx^2 \quad (13)$$

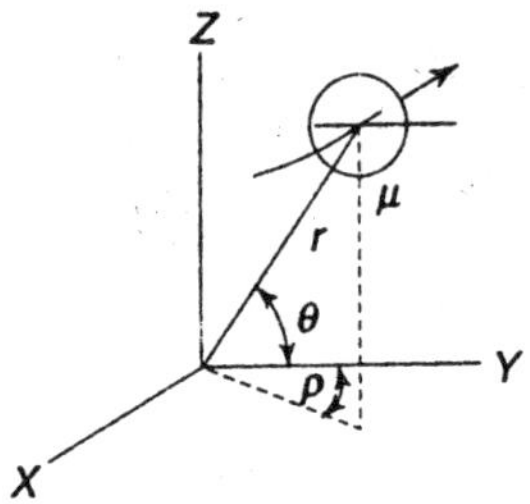

Fig. 1.2

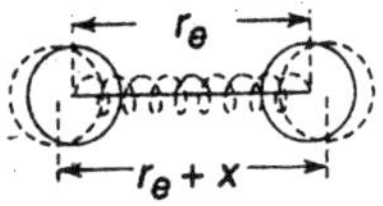

Fig. 1.3

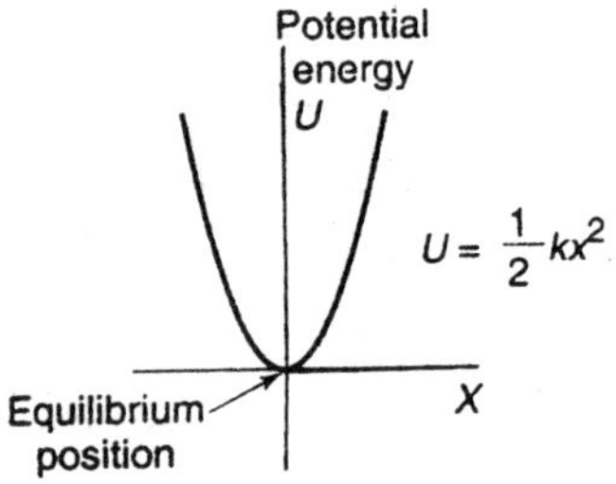

Fig. 1.4

This equation gives a parabola starting from rest at any position x

$$f = \frac{1}{2}kx^2 \quad ...(14)$$

at x = 0, kinetic energy is maximum where r is the equilibrium interatomic distance.

The total E_{vib} at any time is constant.

$$E_{vib} = \frac{1}{2}\mu \left(\frac{dx}{dt}\right)^2 + \frac{1}{2}Kx^2 \quad ...(15)$$

where μ is the reduce mass $\frac{m_1 m_2}{m_1 + m_2}$

E_{vib} for each degree of freedom = kT.

For a diatomic molecule, total average energy per mole is

$$\frac{7}{2}\ RT = (3/2RT + RT + RT).$$

Mechanical Energy

Energy, whether potential or kinetic, that is associated with relatively large masses as compared with the ultimate particles of which the matter is composed, is generally referred to as *mechanical energy.* Mechanical work is expressed in foot lbs (distance times force) cft atmosphere, litres atmosphere, joules. Mechanical power is expressed ft.lb-/min., litres atm per hour watts. A horse power (H.P.) is 33,000 ft. lb per minute.

$$dw = mgh \qquad ...(16)$$

where w is the work done on a particle of mass m through a vertical distance h.

Heat Energy

It was J. Black who pointed out the distinction between the intensive factor temperature and the extensive factor quantity of heat. Equilibrium requires an equality of heat in different bodies.

Energy associated with the motion or the relative position of molecules is called thermal or heat energy. It is a measure of the change in the internal energy of a substance which is due to the rapid movement of particles due to temperature difference.

It can be transferred from one body to another either by radiation or by conduction. The amount of heat required to cause the temperature of one gram of water from 15° to 16° is called a calorie. The amount of heat required to raise the temperature of 453.59 g of water by 1° C is known as C.H.U. and by 1°F is known as B.T.U. 1g cal per g = 1 C.H.U. per 453.59 g = 1 B.T.U. per 453.59 g 1 British Thermal Unit (B.T.U.) is the amount of heat required to raise the temperature of one pound (lb) of water by 1° F.

1 C.H.U. is the amount of heat required to raise the temperature of 453.59 gram of water by 1°C.

1 calorie = 4.1840×10^7 ergs.

B.T.U = British Thermal Unit and, C.H.U = Centigrade Heat Unit.

Radiation Energy

Radiant energy consists of regular radiation impulses travelling outward from electro-magnetic disturbances. The distance between the individual radiant impulses is called the wavelength of the radiations. Table 1 gives some common kind of electromagnetic radiations.

Table 1

Name	*Wavelength*	*Frequency Impulses per sec*	*Source*
Wireless, radio, T.v.	2×10^6 cm	1.5×10^4	Oscillating electric
Short Hertzian,	4×10^3 cm	7.5×10^6	current
radar	3.5×10^{-2} cm	9×10^{11}	
Infrared, or heat	7×10^{-5} cm	4.3×10^{14}	Vibrating molecules and atoms
Visible light	4.1×10^{-5} cm	7.3×10^{14}	
Ultraviolet	1.2×10^{-6} cm	2.5×10^{16}	Vibrating electrons in
Extreme ultraviolet	5×10^{-7} cm	6×10^{16}	atoms
X-rays	1.2×10^{-9} cm	2.5×10^{19}	
Gamma rays	1.10^{-12} cm	3×10^{22}	Nuclear changes
Cosmic radiation	1×10^{-13} cm	3×10^{23}	Nuclear changes

44.7 × 103 kg of gasoline and 157 × 103 kg of oxygen burn to give 400, 000, 000 000 cals, the loss in mass is only 0.0185 g. Similarly, a 1.524 m diameter ball weighing 14,500,000 g would lose only 0.005 g of mass in cooling from red heat to room temperature. At Hiroshima 1 g of matter was converted into energy which produced a blast equivalent to that produced by the explosion of 20,320 × 10^3 kg of T.N.T.

The energy in coal, gasoline, food, automobile battery is referred to as potential energy but it would be appropriate to call it chemical energy. Chemical energy is transformed into other kinds of energy where matter undergoes the proper kind of change, when chemical reactions take place, potential energy is stored or released. The exact form in which this energy is stored is not known in detail.

It must be related to the configuration of atoms and electrons in the molecule. When coal or gasoline burn, chemical energy is converted into heat or light or both. Radiant energy from sun when converted in a growing plant into chemical energy is utilized in making up the stalk, corn, grain and other parts of the plant.

Energy associated with electromotive force (E.M.F.) is called *electrical energy* and generally results from movements of the relatively mobile electrons that are present in varying degrees in all substance. Chemical reactions in various electrical cells are utilized to generate the current and *vice-versa*. Electrical energy is utilized in bringing about a chemical reaction, *e.g.*, in the electrolysis of water, etc. electrical work is expressed in volt coulombs (Joules). Electrical power is expressed in volt coulombs per second or volt amperes called watts.

Atomic, nuclear energy is a form of potential energy associated with the manner in which atoms are constructed. This energy can now be transformed into heat, light, and other types of energy.

TEMPERATURE

It is probably the most important single variable used in science. It is related to kinetic energy of the molecules, *i.e.*, it is a measure of the degree of movement of molecules, atoms or ions.

$$PV = \frac{1}{3} nmv^2$$

$$T = \frac{PV}{nR} = \frac{2}{3R} K.E. \qquad ...(1)$$

To establish a numerical scale of temperature it is essential to define the length of a degree. For this purpose the freezing and boiling points of pure water at the standard atmospheric pressure are taken as the standards and the fundamental intervals arbitrarily divided by 100 in the case of centigrade scale and 180 in the case of Fahrenheit scale. Mathematically in the case of centigrade scale :

$$T = \frac{100(G-G_0)}{G_{100} - G_0} \qquad ...(2)$$

where T is the temperature in centigrades scale and G.G_0 & G_{100} are values of any given property at T, 0 & 100°C respectively. The properties most commonly used are :

(a) Volume expansion of gases, liquids & solids at constant pressure

(b) Pressure increase of liquid and gases at constant volume

(c) Vapour pressure of liquids

(d) Electrical resistance of metals, *e.g.*, platinum

(e) E.M.F. at the junction of two dissimilar metals

The Absolute Scale

A scale independent of the properties of any particular substance is called an absolute scale. An ideal gas is used as the thermometric fluid, *e.g.*, helium, nitrogen or hydrogen. The highest temperatures are those in arcs lying between 3800°C and 10,000°C. The lower limit is absolute zero 273.16°C below the ice point and it has been possible to measure 2×10^{-50} K. The kinetic energy of all molecules or ions or atoms in a perfect crystal becomes zero at absolute zero. There are two important scale known at Kelvin and Rankine scale. 273°C is known as T°K and 460°F is known as T°R.

PRESSURE

The second most important variable is pressure. It is defined as force per unit area. It is expressed in dynes per sq. centimeter or g per sq. inch. Another important unit is atmosphere, *i.e.*, the pressure caused by the weight of earth's atmosphere. It is measured in terms of the height of a column of fluid that the pressure will balance. Mathematically *phd* where p is the pressure, h the height of the fluid and d is the density of the fluid. A standard atmosphere is defined as the pressure due to a vertical column 760 mm long of pure mercury having a density of 13.591 gms per cc at 0°C and in location at sea level at 45° latitude where g is 980.665 cm per sec^2.

In engineering, work pressures are measured above a datum established by the prevailing atmospheric pressure and such pressures are known as *gauge pressures.* Absolute pressure is the sum of the pressure as read by gauge pen to the atmosphere and the barometric pressure at the point. A space from which all the gas is removed would be at zero pressure. The most efficient pump creates a vacuum as low as 10^{-9}mm, but even at this pressure at 0°C there are 2×10^{-30} gm per cc, *i.e.*, about one molecule of H_2 per 100 cft.

Pressure up to 1000 atm are used industrially. In common artillery pressure of the order of 50, 000 lbs are generated. The unit of pressure is torr particularly in expressing low pressure. The torr is equal to 1mm of Hg pressure of 133.322 n M^{-2}. In a uniform gravitation field the potential energy of a particle is given by

$$U = mg\,h \qquad ...(1)$$

where m is mass of the particle and h is the height above ground level. U at the surface of the earth is taken as zero.

According to Boltzmann distribution formula :

$$\left(\frac{n}{v_2}\right) = \left(\frac{n}{v_1}\right) \exp\left(-\frac{U_2 - U_1}{KT}\right) \qquad ...(2)$$

or $$\left(\frac{P}{KT}\right)_h = \left(\frac{P}{KT}\right)_0 \exp - \frac{mgh}{KT}$$

or $$Ph = P0 \exp - \frac{wgh}{RT} \qquad ...(3)$$

$$\left(\text{because } \frac{n}{V} = \frac{P}{KT}\right)$$

This is known as *barometric formula.* Fig. 1.5 gives the variation in atmospheric pressure with altitude.

STATES OF MATTER

Substance exist in three aggregation states namely gas, liquid and solid. These states differ in the kind of motion and in the packing of atoms or molecules. Gaseous state possesses molecules which have translational, rotational and vibrational motion. The distance between the molecules in a gaseous state above its critical temperature is very large, *i.e.*, the molecules are very loosely packed.

In the solid state, the distances between molecules are small , *i.e.*, they are closely packed. The molecules have practically no translational or rotational motion but they do vibrate about fixed centres of equilibrium at a frequency of $10^{13} - 10^{14}$ vibrations per second.

Very low mobility of the molecules or atoms accounts for the resistance of solid to change of shape. The liquid state is intermediate between the gaseous and solid states.

Liquids approach gases in molecular motion and solids in packing density. They have considerable mobility and their molecules continually change their position by changing their centres of equilibrium. This also explains why they change their shape and flow under the influence of very small stresses. In liquids and gases diffusion occurs by group mechanism, *i.e.*, the molecules move jointly with adjacent molecules which can simultaneously move apart making place for other molecules. In solids diffusion takes place only be elementary jumping of atoms to neighbouring unoccupied sites which are always present there due to imperfection in packing. This explains a very slow diffusion amongst solids.

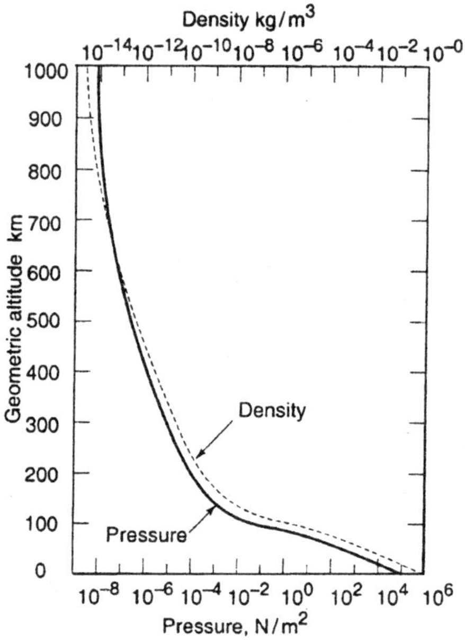

Fig. 1.5

Molecular packing is about the same in the liquid and solid state but it is much different in gases at NTP. The density of a crystalline substances on melting changes by a few percentage but on condensation of vapours it increases by thousand times.

It is the close packing which brings liquid very near to solid and distinguishes it sharply from gases. There is a strong molecular attraction in liquids because of this close packing.

Phase Transitions

Phase transitions are brought about by a change in mutual arrangement of molecules and thermodynamic properties of the substance. Phase transitions are either of first order or second order.

First Order Phase Transitions

These transitions are accompanied by a discontinuous change of internal energy and specific volume. These transitions involves absorption or generation of heat in phenomenon like melting, vapourizaiton, sublimation and many other transformation from one solid state to other solid state. During a first order transition the thermodynamic potential varies continuously leading to a jump in entropy of the system.

Second Order Transitions

These transition involve smooth change in the internal energy and specific volume of the system. No heat is absorbed or evolved, but eh heat capacity and compressibility of the system changes abruptly near the transition, *e.g.*, transition of liquid helium (I) to liquid helium (II), iron to paramagnetic state at curie point, and solid S_8 (Rhombic) to S_8 (monoclinic).

The usual change of state (solid to liquid; liquid to vapour) is called a first order transition. The free energy of the two forms, at the transformation temperature under constant pressure, is equal but there is an abrupt discontinuous change in the slope of energy vs temperature curve. There is a two-phase equilibrium, the upper and lower limits of the value at the discontinuity represents the coexisting phases.

A great many liquids on supercooling below their freezing points display a discontinuity in the temperature derivatives of energy (E), enthalpy (H), entropy (S), volume (V) whereas the quantities E, H, S, V themselves are continuous. These discontinuities are smaller below the transformation temperature. These are known as second order transition. It is in this region of transformation temperature that the supercooled liquids are transformed from viscous liquids into hard and brittle but still amorphous solids designated as glass, *e.g.*, liquid helium.

SOLIDE PHASE STATE

Solids under ordinary conditions have definite volume and shape and tend to maintain these even under deforming forces. Liquids also have definite volume but take the shape of the vessel into while they are poured. Both the liquids and solids are nearly incompressible. Gases, on the otherhand, maintain neither the volume nor shape and completely fill the container into which they are introduced. Gases can be expanded or compressed very easily.

The three states of matter are interconvertible. This can be done by heating or cooling. Heating increases the interparticle spacing and the kinetic energy of the particles. So a solid on interparticle spacing and the kinetic energy of the particles. Many properties of solids, liquids and gases can be easily observed with the help of our sense organs.

The properties which can be observed with the help of our sense organs are called macroscopic properties. The description of the behaviour of the three states of matter an terms of atomic theory is called microscopic description of matter. In chemistry, we try to explain the macroscopic behaviour of matter in terms of its microscopic description. In this chapter, we would try to explain the observable properties of different states of matter in terms of the behaviour of the constituent particles in term.

In a solid, the constituent particles atms, ions or molecules are most closely packed and have the strongest inter molecular force of attraction. As a result particles are fixed in their position and do not have any freedom of motion except that of vibration about their mean position. This give rise to the following characteristic properties to a solid.

(i) A definite shape and definite volume

(ii) Incompressibility

(iii) Rigidity

(iv) No fluidity

(v) Poor diffusibility through them

SOLIDS CHARACTERISTIC

Some characteristic properties of solids are given as follows :

(i) Solids maintain their volume independent of the size of shape of the container in which they are placed.

(ii) Solids are rigid and have definite shape.

(iii) Solids diffuse very slowly compared to liquids or gases. This is because the constituent particles in solids are closely paced and there is very little space to permit the movement of the particles.

(iv) Solids re nearly incompressible. The compressibility of solids is about 10th times that of the gases.

(v) Most solids melt on heating. Some undergo sublimation upon beating. Melting temperatures vary considerably from solids to solid.

(vi) Solids have very high mass-to-volume ratio (density) as compared to gases.

CLASSIFICATION OF SOLIDS

Polycrystalline Solids

In certain crystalline solids, the crystals are very find. These cannot be seen by the naked eyes. Such solids given an impression of being amorphous. Such fine crystalline solids, which appear as amorphous are termed as polycrystalline solids. Metal powders are polycrystalline in nature.

In a polycrystalline sample, there are small crystals. These crystals are randomly oriented. As a result, a sample of polycrystalline material appears to be isotropic, even when each individual crystal is anisotropic.

Amorphous Solids

Amorphous solids have some mechanical properties which are commonly associated with the word solid. But, the amorphous solids differ from the crystalline solids in many respects. So, amorphous solids are sometimes called as pseudo solids.

Some characteristic properties of the amorphous solids are;

(i) The mechanical, electrical and optical properties of amorphous solids do not depend upon the direction, *i.e.*, amorphous solids are isotropic. In this respect, the amorphous solids resemble liquids. It is for this reason that the amorphous solids are considered as supercooled liquids. There is only a short-range order in amorphous solids.

(ii) Amorphous solids do not poses sharp melting points. This indicates the absence of long-range order Amorphous soften on heating and gradually begin to flow like liquids.

(iii) Amorphous solids do not occur in characteristic geometrical shapes.

(iv) When hammered, amorphous material break in an irregular manner. When you hammer a glass sheet or a plastic sheet, these get broken in a very irregular way along many directions.

Examples. Some typical amorphous materials are Plastics, Glass, Rubber, Lamp black, pitch etc.

The solid state is next to gaseous state in its simplicity to understand, although it is also not free from complex problems. Solids are obtained by cooling liquids. There are inherent two possible ways of passing from the equilibrium liquid state into solid state namely crystallization and glass transition.

Crystallization is a transition from a state of short range order to one of long range order in the process of formation of a new phase. The concept of short range order and long range order is defined by the ratio between the distance over which the order extends and the dimensions of the elements arranged in the order.

This is brought about by slowly cooling the liquid where the molecules, due to enough thermal energy and low viscosity of the medium, get enough opportunity to arrange themselves in a geometric order. With the slow release of energy, they grow on the nucleus maintaining the shape of the crystal. It is a phase transition of the first order.

Glass transition does not involve a change of phase, retaining a short range order. It is actually a super-cooled liquid in a non-equilibrium state, which solidify on cooling passing into glassy state. It loses the property of liquid state and acquires the properties of solid state. These changes do not occur abruptly but gradually over a certain range of temperature. The effect of heating an amorphous and crystalline solid. It can be seen that crystals have a sharp melting point while amorphous solids soften over a range of temperature.

Crystalline Solids

All crystalline solids have the following characteristics.

(i) The constituent particles in crystals are generally held by strong interatomic interionic or intermolecular forces. Particles in a solid do not possess translation motion, but they vibrate about their equilibrium positions.

(ii) Crystalline solids possess characteristic geometrical shapes.

(iii) When cut or hammered gently, crystalline solids show a clean cleavage, *i.e.*, they show fracture along a smooth surface.

(iv) Crystalline solids have sharp melting points. This indicate that there exists a long range order in crystalline solids. The long range order in crystalline solids is due to regular arrangement of the constituent particles (atoms, ions or molecules) in three dimensions.

Polymer Crystals

Polymers are products of high molecular mass and their decomposition temperatures are far below their boiling points. Hence, they exist only in liquid or solid state. Although, most of the polymers are amorphous, but the crystalline polymers known as isotactic polymers have high melting point and produce fibres of high tensile strength and excellent mechanical properties. A polymer has two types of structural elements viz. monomer units and chains. The long range order is a prerequisite for the crystallization of polymer, not only in the arrangement of units but of chains as well as in three dimensions. This can happen in three ways:

1. Long-range order of arrangement of both chains and monomeric units in three dimensions.
2. Long-range order of arrangement of chains, but the monomeric units are arranged disorderly.
3. Long-range order of the arrangement of units but short range order of chain arrangement.

The high degree of ordering can be achieved in polymers in two ways, *i.e.*, by crystallization or by chain orientation without orientation of monomeric units. These correspond to crystalline and amorphous (glassy) state. It is possible to have a perfect short-range order in chains but the long-range order in arrangement of chains and monomeric units is not very perfect. Therefore, the examples of well ordered liquids and defective crystals are well known.

Crystallization of Polymers

Many polymers are incapable of crystallization under any condition. Their crystallizability is determined by their chemical constitution, *i.e.*, their packing, energy of intermolecular reaction, chain regularity, flexibility, etc.,

Regularity of Polymer Chain

A long-range order must exist in the chain in three dimensions. Sometimes irregularity in the chain may not obstruct crystallization but random and atactic copolymers cannot be crystallized under any circumstances. Natta by using certain catalysts succeeded in synthesizing several highly crystalline vinyl polymers, *e.g.*, polypropylene, polybutylene and polystyrene. A mixture of the polymers of various degree of orientation was formed and crystalline fraction was separated from amorphous fraction by means of solvents as crystalline polymers are less soluble than amorphous polymers.

Second Order Transition Temperature

A first order transition is one at which a definite change in volume occurs. A second order transition is one at which a change occurs only in the rate of change of volume with temperature. Below T_g, polymers are rigid, glass like material and above T_g they are soft elastometric materials. The second order transitions are caused by thermal agitation releasing portions of the resin chain to rotate around its bonds.

These second order transitions can be measured conveniently by deflection temperature DT tests.

Flexibility of Polymer Chain

Polymers with sufficiently flexible chain can crystallize, but liquid chain polymers crystallize with difficulty because the thermal motion of the momeric, units disturb the established order and the crystal breaks. To crystallize a polymer, it is cooled to a temperature where the motion of its units does not hinder their orientation but excessive cooling may also hinder the rearrangement of the units. Crystallization of each polymer is possible only within a definite temperature interval, *i.e.*, between the glass temperature and flow temperature, *e.g.*, the glass temperature of isotactic polystyrene is about 100°C. Hence, it will only crystallize between 100°-220°C which is its melt temperature.

Packing of Molecules

Polymers obey the principle of close packing and macromolecules must be packed as closely as possible in the crystalline lattice. There are several possibilities for the formation of close packings of polymer chains.

The first crystalline structure is built by close packing of spheres which is exhibited by globular proteins. The formation of such crystalline structures are possible because all the spheres are equal in size, *e.g.*, edestion catalase, iodobenzoyl glycogen.

The second possibility of close packing is the packing of helical macromolecules. The convexities of one helix fit into the concavity of other, *e.g.*, amylose, poly (hexamethylure adipamide), polypeptides.

The third is the packing of long extended chains, *e.g.*, polyamide, polyurethanes, etc. The essential requirement is that the side chain substituents must not hinder regular arrangement of adjacent chains.

Close packing of chain molecules can only be accomplished with flexible chains capable of shifting in parts. Introduction of polar groups gives rise to two opposite effects. On one hand, intramolecular attraction increases, favouring close packing, while on the other hand, the thermodynamic flexibility (potential energy difference between two positions of the chain) of the chain decreases.

The chain becomes more rigid. The rate of close packing depends on these two factors. However, with polar groups like –OH or –CN, the attraction effect prevails inspite the fact that the chains are rigid. Since rigid chains closely pack in the extended state, such polymers are highly oriented, *e.g.*, poly (vinylalchol), polyacrylo butyrate. The chains of isotactic polypropylene are highly flexible and form crystalline lattice, but rigid cellulose chains cannot form such lattices under any condition.

Polymer Crystal Cells

The crystalline cell of polymers consists of parts of chains located at definite distances a, b, and c and angles α, β and γ. They do not differ at all from cells formed by low molecular weight compounds and obey all the rules of symmetry.

Polyethylene crystals form orthorhombic unit cells with a = 7.40Å, b = 4.93Å and c = 2.53Å and density equal to 1.0 g/cm^3. All the carbon atoms in the polyethylene cell are arranged on the same plane forming a zigzag plane. In the polyethylene unit cell, chains are arranged along four edges and in the centre of the cell.

In the case of chains having bulky side substituents helical crystals are formed, *e.g.*, polypeptides, polypropylene divinyl and olefin polymers. One helical loop may contain a different number of monomeric units depending on the nature of a polymer. Natta synthesized 1,2 polybutadiene and similar results were experienced with isotactic polystryrine. Polymers like sulphur or phosphorus are characterized by poly-morphism. Polypropylene may form crystals belonging to monoclinic, hexagonal and triclinic types of symmetry.

They are typical first order phase transitions. These transitions between various polymorphous formations take place either upon a change in temperature or under the action of load which brings a change in the parameters of a crystalline cell. As a matter of fact most of the polymers can be crystallized under stretched conditions. The morphology of a crystalline polymer is determined by the mutual arrangement of the unit cells within the limits of the crystalline state of a body.

Single Crystals

Single or ideal crystals are formed by the parallel transfer of unit cells along edges at distances equal to periods. Fischer first of all ascertained the possibility of high molecular compounds to form monocrystals. Polymer single crystals like other crystals have been found to contain certain defects originating in the packing of chains.

Lamellar Crystals

Polymer crystals are grown by crystallizing from a one per cent solution of polymer under gradual cooling or isothermal aging under the temperature of dilution. The dimensions, shape and regularity of structure of a single crystal depend on the chemical structure of a chain and the physical constraints like temperature, concentration of solution, the cooling rate, the nature of the solvent, etc. The polymer single crystals are monolayer flat plane Lamella which are often rhomboid of 100Å thick and the size of the plate up 1 mp.

The axes a and & of a crystalline cell correspond to long and short diagonal of the rhomboid, while axis c along which the polymer chains are directed is perpendicular to the crystal plane. Since the length of the polymer is very big while thickness of the crystal is of the order of about 100Å, the long chain can only fit in the crystal by turning on its surface by 180°. This is known as chain fold conformation. To have an idea in a crystal of 120Å thickness, the fold shall contain 100 carbon atoms. A molecule having a weight of 105 shall fold about 70 times.

The thickness of a lamella, the dimensions of crystallite and degree of defects of the boundary layers of a crystal are affected by the crystallization conditions. Crystals with folded chain conformation can be obtained when super cooling is in the range 15° – 20°C. Crystallization under high hydrostatic pressure 5-10 × 10^3 atm give extended chain crystals. Polyethylene annealed under a pressure of 700 atm give extended chain crystals. When the melt is cooled under great deformation crystals of different morphological forms can be obtained. Polyhedrons known as hedrites and axialites or avoids are obtained by raising the concentration of the solutions.

Fibrillar Crystals

Fibrillar crystals are obtained by the high rate of evaporation of a solvent from a moderately concentrated solution or at cooling of melt. There are two views about their formation. One section believes that aggregation of rolled lamella is responsible for them while other hold that lamella degenerate in the formation of fibrillar crystals.

In stereospecific polymerization, fibrillar crystals are directly found. The formation is determined by the ratio of the rates of growth of a chain and its crystallization. By changing this ratio crystal either with extended or folded chains are obtained.

Globular Crystals

In globular crystals, macromolecules in coiled conformation form the lattice points. These are found in biopolymers, as a very high degree of uniformity of macromolecules is a pre-requisite for it, *e.g.*, tobacco mosaic virus synthetic polymers do not form them.

Spherulites

Spherulites are obtained when polymers crystallize from melts or concentrated solutions. They are made of crystallites which grow radially from one common centre. Macromolecular chains make at least an angle of 60° with the radius, *i.e.*, they are arranged tangentially to the spherulite radius Spherulites possess anisotropic properties. Their indices of refraction of light are different in radial or tangential directions. The orientation of the crystallographic axes continuously changes along the angular coordinates, A spherulite is called positive when the refractive index in the radial direction is greater than in the tangential one; otherwise it will be a negative spherulite.

In radial Spherulites one of the axis of the crystalline lattice retains a constant direction along all the radial directions. Ring type Spherulites are built of lamellar crystals whose orientation continuously changes along the spherulite radius.

Depending on the degree of cooling, the same polymer may form different types of spherulitis. With low degrees of super cooling ring type Spherulites are obtained while radial Spherulites are obtained by high ones.

The formation of Spherulites passes through several stages. First the crystal nuclei are formed which are scattered throughout the volume of the sample and then independent growth of discrete fibrillar or lamellar crystalline structure grow at the same time in all directions. Then encountering one another they grow into regions filling the entire volumes of the body. Then their boundaries are distorted and they assume the form of a polyhedron.

The structural units in spherulite are bundles of macromolecules made out of chains in an extended conformation. This is possible when several macromolecules try to crystallize simultaneously from both the ends. Such interstructural bonds combine to form separate crystallites inside fibrillar or lamellar crystals.

Oriented State of Polymers

Polymers because of great anisotropy of the dimensions of macromolecules exist in oriented state. In the extreme case all the macromolecules will be oriented parallel to each other, but in reality this can happen on the average as the flexible chains do retain some fold conformations. In the oriented state the difference in amorphous and crystalline polymer vanishes.

Peterlin suggested that in oriented polymers microfibrils of diameter 100-200Å clearly divide crystalline blocks of chains with fold conformation. These chains are divided by defective layers and a large number of intra- and smaller number of interfibrillar chains. The microfibrils continue into larger fibrils to create the macrostructure of oriented bodies of the size 1000Å. These polymers are important for producing the fibres and films. In the first case an uniaxial orientation takes place while in the second a planar orientation takes place.

Liquid Crystals

Certain substances that contain rod-shaped molecules can exist between the crystalline and liquid state. The forces between these molecules are greatest when they-lie parallel and the crystals of such substances consist of parallel arrays of molecules stacked side by side when the crystal is heated, a point is reached at which the thermal energy is sufficient to break down the crystal lattice but not the parallel arrangement of the molecules.

This results into a liquid crystal when the molecules are free to move provided they all remain parallel. At higher temperature the liquid crystal melts to a liquid. Such crystals are called *nematic*. There are other liquid crystals where the molecules are not only parallel but their ends are aligned. Such crystals are known as *sematic*. In this case the crystals contain layers where the parallel molecules are free to move about in its layer but not in other layers. These crystals form drops in which the surface forms sharp steps with flat risers in between.

The intermolecular attraction in liquid crystals do not act equally in all directions with the result that all the degrees of freedom are not liberated. In a nematic liquid crystal the molecule can rotate only about its long axis and two rotational degrees of freedom are lost. In sematic liquid crystal the molecules are free to move only in the two directions and some translation degree of freedom is lost.

LIQUID PHASE STATE

Since the molecules are adjacent to one another, the question of random packing does not arise but they maintain a short range order, *i.e.*, the order is observed over distances commensurable with molecular size. In the vicinity of any given molecule its neighbours may be in definite order but at a small distance from the molecule there is no order. Short range order in liquids can be detected by X-ray studies and is manifested as density fluctuations and in the formation of ordered clusters of molecules called associates. It was considered that these associates represent quasi-crystalline structure but nowadays analogy of liquids is drawn with gas. Which also show diffraction patterns near critical temperature.

The liquid phase is the state of all solid substances (crystalline and amorphous) above their melting point. Since silicate glass has no crystalline lattice, all amorphous solids are said to be glassy or glasses. Glassy and crystalline bodies are both in solid aggregation state and do not differ in mobility or packing of their molecules. There are two phase states in solids namely crystalline and amorphous (glassy). The liquid phase state covers liquid and solid (glassy) above the melting point. The solid state in the sense that there is neither the ordered arrangement of constituents nor the complete disorder as in gases.

It is generally observed that some of the properties of liquids closely resemble those of gases while some of the properties approach those of the solids. In the light of the kinetic molecular theory, liquids may be regarded as a continuation of gases into the region of small volumes and high intermolecular attraction. The cohesive forces in a liquid have been stronger than those in gases even at high pressures, and are sufficiently high to keep the molecules confined to a definite volume. The positions of the molecules in liquids arc not rigidly fixed. These forces are not strong enough to entirely eliminate the movements of the molecules in the liquid. Hence, the molecules in the liquid state have much shorter mean free path as compared to gas molecules.

Properties of liquids can be explained on the basis of their following characteristics :

(a) Appreciable forces of attractions exist between the molecules of a liquid. These are about 10^6 times as strong as in gases. The forces amongst the constituents of a liquid disallow them from separating spontaneously from each other, but they are not strong enough to hold the molecules in fixed positions. It is possible to explain surface tension, viscosity, fairly high heat of vaporization and vapour pressure of liquids in terms of these attractive forces.

(b) The molecules in a liquid are in a state of random motion although the extent of randomness is appreciably much smaller in comparison to gases. The molecules are relatively close together. Most of the space in the liquid gets occupied by its molecules and only a small fraction of the space is available to them for their free movement. This explains the higher density, incompressibility and slow diffusion of liquids in comparison to gases.

(c) The average kinetic energy of the molecules in a liquid is proportional to the absolute temperature. Increase in temperature would increase the proportion of the energized molecules, lowers the attractive forces between the molecules, and consequently increases the vapour pressure of the liquid.

From me above arguments, it is evident that most of the characteristic properties of the liquid arise due to the nature and magnitude of the intermolecular forces between the molecules. The important properties dealt with in this chapter are :

(i) Vapour pressure,

(ii) Surface tension, and

(iii) Viscosity.

LIQUID-VAPOUR EQUILIBRIUM-VAPOUR PRESSURE

The molecules that are having sufficient energy to overcome intermolecular attractions are able to escape from the bulk into the gas phase (or vapour phase) provided they are close to the surface. This happens during evaporation. But the average kinetic energy of molecules in vapour state is more than that in liquid state:

(a) The temperature of the liquid,

(b) Attractive forces in the liquid,

(c) Surface area, and

(d) Pressure above the liquid.

If we examine the evaporation of a liquid in an enclosed space (at constant temperature), it is found that some of the molecules of the liquid will spontaneously pass from the surface into the space. But the space is closed. Therefore, the molecules in the vapour state are unable to escape from the container. The molecules in the vapour phase collide with each other and with the sides of the container. Some of the molecules on collision pass on their energy to other molecules and come back to the liquid phase. This phenomenon is termed as condensation. The rate of condensation of molecules in the vapour phase is proportional to the concentration of molecules in the vapour phase. A stage is ultimately reached when the rate of condensation becomes equal to the rate of evaporation. This situation corresponds to that of a dynamic equilibrium between the liquid and its vapour.

VISCOSITY

Liquids possess another property known as viscosity which implies resistance to flow. It is developed in liquids because of the shearing effect of moving one layer of liquid past another.

Consider a liquid to be divided into parallel planes or layers of molecules, a fixed distance apart. Let the area of each plane be A and the distance between planes be dy. Suppose each plane is moving to the right with velocities v_0, v_1, v_2 etc. such that each succeeding velocity is greater than the preceding by an amount dv.

This type of flow is called *laminar flow* or *streamlined flow as distinct from turbulent or vortex flow* in which parallelism of the layers is not maintained. In laminar flow the force 'f' required to maintain a steady velocity difference 'dv' between any two parallel planes is directly proportional to the area of plane A and the velocity difference between the two planes dv. It is also inversely proportional to the distance between the planes.

Mathematically, it can be expressed as

$$\therefore \qquad f \propto A.\frac{dv}{dy}$$

$$\therefore \qquad f = \eta.A.\frac{dv}{dy} \qquad ...(i)$$

where, η is called the coefficient of viscosity of the liquid.

$$\therefore \qquad \eta = \frac{f}{A \times \frac{dv}{dy}} \qquad ...(ii)$$

When $A = 1$ sq.cm and $dv = 1\text{cm sec}^{-1}$,

$dy = 1\text{cm}$ then η = force in dynes

$$\text{units } \eta = \frac{\text{dynes}}{\text{cm}^2 \times \frac{\text{cm/sec}}{\text{cm}}} = \text{dynes.sec.cm}^{-1}$$

This is the absolute unit of viscosity coefficient. The practical unit in the C.G.S. system is known as poise.

$$\therefore \qquad 1 \text{ poise} \equiv 1 \text{ dynes . sec cm}^{-1}.$$

Other units in use are centipoise (0.01 *poise*) and *millipoise* (0.001 *poise*).

Therefore, the coefficient of viscosity (η) 'Can be defined' as the force in dynes that must be exerted between two parallel planes 1 cm^2 in area and 1 cm apart in order to maintain a laminar velocity difference of 1 cm sec of one layer past another.

METHODS OF DETERMINING VISCOSITY COEFFICIENT

The method commonly used for determining viscosity coefficient is based on Poiseuilles equation,

$$\eta = \frac{\pi P r^4 t}{8LV} \qquad ...(iii)$$

where V = volume (in cc) of a liquid of viscosity η that flows through a capillary tube of radius r cm and length L cm in time t secs under a pressure head of P dynes/cm^2.

Instead of determining η by a direct application of the Poiseuille equation, the one commonly used is the one devised by Ostwald in which the viscosity of one liquid is compared with that of a liquid of known viscosity. (*e.g.*, H_2O).

Use of Ostwald Viscometer

A known volume (say V_0 cc) of the given liquid is introduced into bulb B of the viscometer. The liquid is then gently sucked into the bulb A above the mark 'x' and the time (t_2) required for the liquid to flow through the capillary from 'x' to 'y' is accurately determined using a stop watch. A mean of three observations is recorded. The viscometer is then cleaned and dried and the same volume (V_0 cc) of water, of known viscosity, is introduced into bulb B. This is again sucked into bulb A gently and the time (t_1) required for water to flow between marks 'x' and 'y' is determined as before.

From equation (3), we have,

For water $$\eta_1 = \frac{\pi P_1 . r^4 . t_1}{8LV}$$

For the given liquid $$\eta_2 = \frac{\pi P_2 . r^4 . t_2}{8LV}$$

Note that V = volume corresponding to that between the marks

'x' and 'y'

$$\therefore \quad \frac{\eta_2}{\eta_1} = \frac{P_2 . t_2}{P_1 . t_1} \qquad ...(iv)$$

$$\text{But } P = \frac{\text{Force}}{\text{Area}} = \frac{mg}{A} = \frac{V_0 . \rho g}{A}$$

$$\therefore \quad \frac{P_2}{P_1} = \frac{V_0 . \rho_2 \, g/A}{V_0 . \rho_1 \, g/A} = \frac{\rho_2}{\rho_1}$$

From equation (4), we get

$$\frac{\eta_2}{\eta_1} = \frac{\rho_2 . t_2}{\rho_1 . t_1}$$

$$\therefore \quad \eta_2 = \frac{t_2}{t_1} \times \frac{\rho_2}{\rho_1} \times \eta_1 \qquad ...(v)$$

Since all quantities on the R.H.S. of equation (v) are known, η_2 can be calculated. This is called the *absolute viscosity* of the given liquid. Then η_2/η_1 is the relative viscosity of liquid 2 w.r.t. liquid 1. If liquid 1 is water, then instead of using the term "relative viscosity", for the ratio η_2/η_1 we use the term "*specific viscosity*".

STEAM DISTILLATION

In the case of completely immiscible or almost completely immiscible liquid pairs, the addition of one liquid to the other does not affect the properties of either liquid. Hence, each liquid behaves as if the other is not present. Accordingly, in a mixture of two immiscible liquids, each liquid exerts its own vapour pressure corresponding to the pure liquid at the given temperature. The total vapour pressure P above the mixture is therefore, the sum of the vapour pressures of the two pure liquids at that temperature. Thus,

$$P = P^\circ_A + P^\circ_B$$

Where P°_A and P°_B are the vapour pressures of the two pure liquids A and B at that temperature. The pressure exerted by each layer and hence the total pressure do not depend on the actual or relative amounts of the liquids present.

Any system, as we know boils at a temperature when its total vapour pressure becomes equal to the external pressure. In this case, since the two liquids together can reach any given total pressure at a lower temperature than either liquid alone, it is obvious that the mixture would boil at a temperature than either liquid alone, it is obvious that the mixture would boil at a temperature lower than the boiling point of either of the two liquids. Further, since at any given temperature there is no change in total vapour pressure with change in composition, the boiling point of all possible compositions of any two immiscible liquids remains constant, as long as the two liquids are present. The temperature rises to T_A or T_B depending upon whether A or B remains, only when one of the liquids is boiled away.

The relative proportions of the two liquids in the distillate can be calculated, since the number of moles of each component in the vapour phase is proportional to its vapour pressure. At the boiling point T, if n_A and n_B are the number of moles of the two liquids A and 5 in the vapour phase, then

$$n_A \propto P^\circ_A$$

$$n_B \propto P^\circ_B$$

or $$\frac{n_A}{n_B} = \frac{P^\circ_A}{P^\circ_B}$$

The composition of the vapour remains a constant, since P°_A and P°_B are constants at a given temperature T. If W_A and W_B are the actual weights of the two liquids A and B in the distillate, and M_A and M_B the respective molecular weights, then

$$\frac{P^o{}_A}{P^o{}_B} = \frac{W_A\ M_B}{W_B\ M_A}$$

or

$$\frac{W_A}{W_B} = \frac{P^o{}_A\ M_A}{P^o{}_B\ M_B}$$

The above equation relates directly the ratio of the weights of the two components present in the distillate of a mixture of two immiscible liquids to the molecular weights and vapour pressure of the two pure components. Examples of such pairs include water-cyclohexane, water-nitro-benzene, water-bromobenzene etc., Distillation of immiscible liquids is utilized industrially and in the laboratory as it involves lowering of boiling points of the components. Purification of organic liquids which either have very high boiling point or tend to decompose when heated to their boiling point can be suitably carried out using the above principle.

The other liquid is generally water and the process is referred to as *Steam Distillation.* The immiscible mixture of the liquid and water is either heated directly or by passing the vapours of steam into the liquid, and the vapours distilling over are condensed and separated. In this manner it is possible to distil many organic liquids of high boiling point at temperatures below 373K *i.e.*, the boiling point of water. Steam from the steam generator is passed into the round-bottom flask containing the liquid, say nitrobenzene or bromobenzene to be steam distilled.

The tube carrying the steams dips into the liquid and the flask is heated gently on a sand bath to avoid too much condensation of water into it. The vapours of the organic liquid mixed with steam distils over, condenses and collects in the receiver.

The organic liquid is then separated from the aqueous layer and finally dried. As discussed earlier, the proportion by mass of the organic liquid that distils over is related directly to its vapour pressure and molecular weight. Thus a higher proportion of the liquid is obtained for liquids with high molecular weights and having a relatively high vapour pressure at about the boiling point of water.

TRANSFORMATION TEMPERATURE

The temperature at which an abrupt change in the coefficient of expansion takes place in a glass heated at about 4°C per minute is known as the transformation point Tg. It is a characteristic temperature for a given composition. Since the molecular vibrations just begin at this temperature, it is also known as the lower critical point. Actually, it is a transition region, above which it is fluid or rubbery and below which it is glass. This is because of the interaction between various chains. According to modem point of view, it is a temperature at which configuration becomes comparable, with the rate of approach to equilibrium with the time required to make an observation. The corresponding viscosity is $10^{13} - 10^{14}$ poises.

The transition region in organic glasses is characterized by changes in derivative functions, such as specific heat and coefficient of expansion. Any molecular theory of their properties leads to molecular interpretation of the glass transition, *e.g.*, ΔC_p is close to 11.2 J.K.^{7} per bead where bead is identified as the smallest unit in polymer chain capable of free rotation.

The glass transition temperature is associated with the fractional free volume having a constant value 0.025. Attempts to correlate Tg to polymer structure have focussed on either molecular cohesion or chain stiffness or both. Gibbs and coworkers predicted a second order transition in polymers at a temperature T_2 which is about 60°K below Tg and is nearly equal to the temperature at which the axtrapolated viscosity of the liquid tends to infinity. The molecular motions are found to occur in glass polymers, but in most of the cases it is not clear as to what they are. Therefore, it is safe to assume that glass transformation is associated with large-scale movements.

RELAXATION

The term relaxation expresses the establishment of a static equilibrium in a physical system. In a liquid continuous regrouping of molecular clusters appear under the influence of thermal mobility. Boltzmann's law suggests a method of working out a probability of the process. The reciprocal of this probability of regrouping gives the relaxation time, which is of the order of $10^{-4} - 10^{-6}$ sec in the case of polymers as compared to 10^{-10} sec in the ordinary liquids. Thus, polymer chains possess low mobility and higher relaxation time. The process in which equilibrium is established with time is known as relaxation process.

The curve between temperature and deformation. This consists of three positions, namely glassy state, rubber like state and viscofluid state. The transition from the glassy state to rubber-like state involves a change in the nature of super molecular structure but the phase remains the same as there is no heat effect. The difference between glassy state and the rubber-like state is in the mobility of their macromolecules and super molecular structure. Thus, glass transition is not a phase transition.

MECHANISM OF GLASS TRANSITION

The glass transition depends upon the ratio between the intra- and intermolecular interaction energy and the thermal energy of the motion of their units. The energy of intermolecular attraction does not depend on temperature, but the thermal energy (KT) decreases sharply with decreasing temperature and at a certain definite temperature the thermal energy is insufficient to overcome the energy of interactions. This increases the viscosity of the polymer and decreases the intensity of thermal motion of its units.

METHODS OF DETERMINING T_g

The transition of a polymer from the rubber-like to glass state is accompanied by a gradual change in its physical properties and T_g can be determined by studying the variation of these properties with temperature. The most popular methods are:

(1) specific volume (dialatometric method)

(2) heat capacity

(3) modulus of elasticity

(4) deformation

The difference between a glass and the corresponding liquid or supercooled liquid may be demonstrated by the isobaric change in volume with temperature for an imaginary polymer. The polymer melt on cooling can form (a) a crystal or (b) a glass depending upon experimental conditions. If the substance crystallizes, *i.e.*, follows the path A, B, C, D, it is observed that as the liquid cools from any temperature T, to T_f, there is jump in the volume at T_f, the freezing point. Normally there is marked decrease in volume but in some cases like water, it increases. The coefficient of expansion of the liquid and of the crystal are constant in the vicinity of T_m. As the liquid is cooled slowly, its molecules arrange themselves into a definite pattern forming a crystal. In some of the liquids the forces of attraction (covalent or hydrogen bonds) between the molecules are so strong even above the freezing point, that they cannot freely move and do not provide a suitable crystal nucleus, the crystallization is prevented. When the liquid is rapidly cooled beyond its freezing point T_f the molecules do not have the time to organize themselves all over the liquid and hence, it is supercooled. The individual atoms or molecules are in an ordered array over a short range, but there is no long-range order.

They are indeed in structure, very much like a snapshot of a liquid and are called supercooled liquid. Molecular movement in such liquids is minute and the attempt to describe the flow properties in terms of viscosity to the rigid glass is difficult. Below Tg*l* the dependence is again linear. Te is the start of the solidification and Tg*l* is the termination of the solidification. The temperature at the intersection of the extrapolated curves Tg is known as glass temperature. The molecules in the glassy state oscillate about their equilibrium position and their interchange of places either does not occur at all or only very rarely.

The decrease in the thermal energy of the molecules with the fall in temperature decreases their ability to break down strong covalent bonds, when most of the motion is frozen at a temperature Tg*l*, bearing the liquid in a random and disordered state, a solid state results which is known as glassy state. One may say that the glass represent the metastable state of order of a liquid which is frozen in the range Te & Tg*l* and preserved at the lower temperature rather than a specific kind of material.

For the supercooled liquid in the range Tg and Te at a given rate of cooling (glass 1 A 2) the equilibrium distribution corresponding to each temperature is reached within the duration of the measurement, since the substance is in thermodynamical equilibrium with reference to arrangement and distribution of molecules. When a liquid on cooling changes to crystal, there is an evolution of latent heat associated with the abrupt change in the disorder, but there is no evolution of latent heat during the liquid to glass transformation. This suggests that the molecular arrangement in glass is similar to that in liquid. The glassy state is obtained by the inhibition of a kinetic process rather than a second order transition.

STRUCTURE

The structural units in glass are randomly linked and the regularity in the lattice has been found to extend to 5–10° A only. So it is evident that it has only a short-range order. Silicates are known to exist in three crystalline and one amorphous form as shown below:

$$\text{Quartz} \xrightarrow{870^o} \text{Tridymite} \xrightarrow{1470^o} \text{Crystabollite} \xrightarrow{1710^o} \text{glass}$$

Silicon atoms are bound to four oxygen atoms forming a regular tetrahedron. The distance between the centres of the silica and oxygen ions is 1.6° A. In crystoballite the structure consist of a set of interlocking rings each with six atoms, three silicon and three oxygen held together by covalent bonds forming a giant molecule like that of carbon in diamond. All the three varieties soften at 1600° and melt to a liquid at 1700°. The silica crystal can melt only by breaking of the bonds. In the liquid a large number of Si-O bonds have broken to allow the parts to move about.

However, the breaking is not irrevocable; Si-O bonds are continuously forming and breaking in the melt so that most of the time silicon is attached to four oxygen atoms and oxygen to silicon atoms. When the melt is cooled rapidly, the particular bond pattern present at that instant is frozen to a liquid of very high viscosity. Instead of a regular pattern of six membered rings, the tetrahedron is distorted resulting in jumbled rings of different shapes and sizes. The positional disorder existing within each distorted tetrahedron prevents the prediction and assignment of fixed position to the ions.

The resulting glass is still a giant molecule but a disordered one. It is debarred from changing into a regular structure since that would involve the breaking and reforming of strong covalent bonds. The strong directional covalent bonds, present in the liquid, form wrong connections leading to glass. Although, a short-range order is followed they form rings with the wrong number of elements. Moreover, the atoms in glass are at rest and there is a fairly constant distance between each atom and its -neighbour. The large interatomic distance do not have the same regularity as in the crystals.

Therefore, the glasses have partially ordered material as is evident from two or more diffused haloes in the X-ray photographs. In view of very strong covalent bonds and very high viscosity of the liquids, the probability of the atoms to break their connections and to reform in the right direction is very low. Therefore, they form a disordered arrangement with strong interactions, resulting in glass. Since different energies will be required to breakdown the different meshes of structure, there is a continuous softening range instead of abrupt points.

Warren ruled out the possibility of the existence of molecules in the glass and found very good confirmation in the calculated and observedmaximas. It was very convenient to describe the change in viscosity on cooling to an association or polymerization process.

Zachiarasen deduced the following condition to be followed by an oxide glass of the composition $A_m O_n$: (1) No oxygen atom can be bound to more than two A atoms, (2) The coordination of oxygen around A must be low. (3) The oxygen polyhedra must not share edges or faces but only comets, and (4) At least three comers of the oxygen polyhedra must be shared. There are two types of substances that follow the above conditions: Those substances, that have directed covalent bonds, crystallize in an open structure, *e.g.*, SiO_2, B_2O_3, GeO_2, P_2O_2, As_2O_5, V_2O_5, Sb_2O_3, Sb_2O_5, As_2O_3, ClO_5, Ta_2C_5. Substances whose liquids consist of long molecules which become entangled on cooling and require breaking of many covalent bonds for rearrangement, *e.g.*, plastic sulphur, organic glasses.

ORGANIC GLASSES

It is perhaps not always recognized that glass formation is a property shared by a very large class of compounds. These are polyphosphate glasses, those formed by certain elements in the polymeric form, *e.g.*, S, Se, those formed by hydrogen bonded materials like alcohols and organic polymers.

Lourenco obtained a highly polymerized viscous undistillable residue by polymerizing ethylene glycol in presence of ethylene dihalide, but Staudinger was the first who suggested chain formulas for organic polymers. The polymeric glasses comprises a class of plastic materials of long chain organic molecules having an amorphous structure. The polymer chains consist of single repeating monomer units joined together by covalent bonds. Polanyi and Herzon suggested that no unit cell of crystal can be smaller than a single molecule. In 1929 Staudinger clarified the confusion regarding the structure of polymers.

However, certain plastics are also used as copolymers, *i.e.*, polymers containing more than one monomer unit, *e.g.*, styreneacrylonitrite copolymers of vinyl chloride, vinylacetate and nylon, synthesized from adipic acid hexamethylene diamine with an elementary unit—$NHCO(CH_2)_4CONH(CH_2)_6NH$—of a length of 17°A, which on melting at 260°C gives a true glass. Carothers and associates made an effort to link physical properties, *e.g.*, viscosity, double refraction, rubber elasticity, etc., with molecular weight and shapes of these molecules. Flory has worked out several methods to determine the various physical properties of the polymers. Zachariasen propagated the theory of random continuous network for polymers.

In the case of linear polymers each chain molecule retains only two functional end groups during the process of polymerization irrespective of its size. The main chemical type of polymer glass may be represented by polystyrene, polymethyl methacrylate, polyvinylchloride, and polycarbonates, etc. Thus, polystyrene normally contains a small quantity of lubricant which reduces the glass transformation temperature by 10°C and improves its moulding properties.

In nonlinear polymers the number of functional groups increase for each molecule with its size or us molecular weight, *e.g.*, formation of various sulphur, where S_8 ring opens and produces a chain between two molecules by the pairing of electrons.

Polymethyl methacrylate is a glass below 90°C. The mixture of polymer and monomer, or of polymer, monomer and the diluent, transform to a glass. During the process of polymerization, the percentage of polymer reaches a certain well defined high value. Even the monomer molecules are frozen up in their position in the glassy state.

Recently, the organosilicone compounds, because of their greater stability, have become important. The Si-CH_3 or Si-C_2H_5 bonds have greater oxidation resistance than the corresponding C-C bonds. Depending upon the molecular weight, chain length, ring structure and the degree of three-dimensional polymerization, the methyl or ethylsilicones can have very different properties ranging from fluid to viscous oil, resins and elastomers. The products obtained depend on the methods of preparation.

Warren, Bagchi, Alexander and Miller, have discussed and given upto date account of the application of X-ray to glassy materials. Debye, Station and Heikens have used the X-ray scattering method for determining the void spaces in polymers and glasses. Interesting relationships have been observed between the heterogeneity of the glass and mechanical properties of fibres. Several workers have also applied NMR, ultrasonic and diffusion techniques for the above purposes.

The constitution of an organic polymer is guided by the following parameters:

1. The configuration describes the atomic structure of the molecule according to the chemical formula. The configuration is the same in solution, in the glassy state or the crystalline state.

2. The conformation is a dynamic parameter, *e.g.*, the rotation of the segments about a single bond. They may have different number of elementary units per turn of the helix but in every case the number of possible conformation will be greater in liquid or in the glassy state than in the crystal.

3. The molecular packing or chain packing refers to the lateral organization of the chains in the crystal. Since most of the polymers even as crystals are only poorly defined, this parameter covers different types of disorders and lattice imperfections.

Du Pont have reported new polymers called ionomers by crosslinking chains of polyethylene through the ionic forces acting between (–COO) groups and Mg^{+2} or Zn^{+2} as cations. Rees and Vaughan have been successful to introduce strong ionic forces between the chains through their carboxylic groups.

The ionomers promise a rare insight into the effect of gradual transition of Van der Waals forces and hydrogen bonding into the three-dimensional ionic network. The ionomer resins are solids with a high vibrational entropy, introduced by the electrical fields, screening demands and nucleating effects of cations which interfere with the establishment of long-range order. All these facts result in the following properties:

1. The stiffness increases with the acid level and the concentration of cations.

2. The ionomer resins exhibit unusual stability in high voltage application.

3. The ionomer resins have unusual elasticity and toughness.

4. The ionomers, because of association in liquid state resulting in high viscosity, behave like high molecular weight thermoplastics than like cross-linked polymers.

5. Compositions containing dibasic acids in conjunction with di- or trivalent ions resemble cross-linked olefins in the behaviour of the melt.

6. The temperature coefficient of viscosity is very sensitive to the strength of bonding forces.

Covalent C-C bond in cross-linked polymers present flow at temperatures at which the polymer is thermally stable. Ionic bonds and resulting association in ionomers makes the viscosity of the ionomer very sensitive to temperature giving activation energy of the order of 20 kcal mole.

Hydrogen bonding through carboxylic acid molecules makes the viscosity less sensitive to temperature where the activation energy is 14 kcal/mole. Van der Waals forces in the polymers show the lowest temperature coefficient, *i.e.*, 7-19 kcal/mole.

The intermolecular binding forces determine its solubility in organic solvents and permeability for oils, lonomers have extremely low permeability for oil and are insoluble in common solvents. Special mixtures, *e.g.*, N, N-dimethylacetamide, $CH_3CON(CH_3)$; added to decalin can dissolve ionomer.

Rose and Block have synthesized some inorganic polymers by refluxing dibutylphosphoric acid with zinc acetate. The above polymer remains flexible down to –80°C. It is soluble in chloroform and a polymer of molecular weight 10,000, melts at 150°C and forms a glass on cooling.

INORGANIC GLASSES

Obsidion, a mineral found in nature, is a dark reddish brown, translucent glass, formed when molten rock is rapidly cooled. These inorganic glasses are the products of fusion of inorganic salts cooled to a rigid mass without crystallization. Winter observed that oxygen, sulphur, selinium, and telurium elements of group VI of the periodic table form glasses with other fifteen elements of group III, IV, and V. These group VI elements are found to form binary glasses and are self vitrifying. A discussion about these is beyond the scope of the present work and interested readers are advised to read the original work. These include various oxides, chlorides, nitrates, sulphates, thiosulphate, alums and other hydrated salts.

DEVITRIFICATION

The molten mixtures of alkali silicates, on cooling slowly, try to crystallize out into one or more of the compounds at their freezing point and the whole mass loses its glassy nature. This phenomenon is known as the devirtrification of glass. All glasses except those having very high viscosity in the molten state try to devitrify. Reamer produced some polycrystalline ceramics by heating glass but it was actually Stookey who systematically studied this subject and actually produced ceramic by heating glass under controlled condition. The small devitrified glass crystal slowly grow into a strong, opaque, white mass, known as porcelain. They have a much higher mechanical strength and markedly improved electrical insulation properties than ordinary glass.

VISCOSITY TEMPERATURE CHARACTERISTICS

Glass on heating softens continuously and progressively to fluid due to continuous fall in viscosity. Gehlhoff and Thomas found a correlation between the change in viscosity and its composition. Therefore, the relationship between viscosity and temperature is a very important parameter for the glass. Important work in this direction is carried out by Field, Royster, Herty and McCaffery. They have made a thorough study of change in viscosity with temperature and with time. Lillie has produced the most accurate results using modified Margules method. The complete viscosity temperature curve. It would be of interest to define the viscosity temperature relationship in terms of certain characteristic temperature at which the glass attains a particular value for viscosity.

Certain temperatures are only of historical importance, *e.g.*, deformation temperature, cohesion temperature, flow temperature, softening point, etc., while others like working point, Mg point, annealing point, strain point and transformation point are very important in the manufacture of glasses.

Working Point

At the usual melting temperatures glass has the same viscosity as glycerine at 0°C, *i.e.*, 10^2 poise. For giving a shape to the glass, it should remain in the fluid condition over a wide range of temperature. Therefore, the glass is removed from the furnace at a viscosity of $10^4 - 10^5$, poise so that it can be given proper shape. The corresponding temperature is known as working or flow temperature.

Softening Point

Littleton defines it as the temperature at which a uniform rod of 0.5 to 1.00 mm diameter and 22.9 cm long elongates under its own weight at a rate of 1 mm per minute under standard conditions. The viscosity at this temperature is $10^6 - 10^7$ poises.

Mg Point

The temperature at which the viscosity flow exactly counteracts the thermal expansion during the measurement is known as dilatometric or

Mg point. According to the recommendations of the Society of Glass Technology it is the maximum point reached on the complete thermal expansion curve for the glass, *i.e.*, the point normally corresponding with the softening temperature. The viscosity at this temperatures is $10^{11} - 10^{12}$ poises. It also corresponds to the extreme upper limit of the annealing range and is denoted by A or the upper critical point designated by C_t. This temperature also depends on the thermal history of the glass.

Annealing Point

Glass on being cooled, without special heat treatment, develops unequal strained condition leading to a tendency to break. To prevent this strain, exact controlled temperature is maintained during a short interval known as annealing range. The temperature at which the internal stress in the glass is relieved within fifteen minutes, is known as annealing point. The viscosity corresponding to this temperature is 10^{13-14} poises.

Strain Point

Littleton defines the lower limit of the annealing point as the strain point. This is also the highest temperature from which the glass can be rapidly cooled without introducing serious internal stresses. The glass can be commercially annealed within sixteen hours. The viscosity at this temperature is 10^{14-16} poises.

GLASS TRANSITION TEMPERATURE AND MOLECULAR MASS

Glass transition increases rapidly with the increasing mass and finally reaches a constant value. The molecular mass at which T_g becomes constant depends on its chain flexibility. For isobutylene T_g becomes constant at M equal to 1000 while polystyrene M should be more than 15,000.

Low molecular weight polymers exist only in two states, *i.e.*, the glass and the liquid and the mass transition temperature coincide with the flow temperature. But, with increasing molecular weight at a certain molecular mass the transition temperature splits into T_g and T_f, thus, introducing a rubber-like state. With further increase in mass, T_g becomes constant but T_f continues to increase. Thus, $T_f - T_g$ gives the temperature range of rubber-like state.

It has been experienced that low molecular mass polymers exist as glassy or liquid where their T_g and flow temperature coincide. With higher molecular mass $T_g - T_g$ is great showing the increase in rubber range. The molecular mass at which the transition temperature splits into T_g and T_f depends on the chain flexibility. The more rigid the chain, higher will be the molecular mass at which the T_g splits. Thus, in polyisobutylene rubber-like strain appears at molecular mass 1000 while in polystyrene, it appears at M 40000.

Polymers with less flexible chains have very high T_g and they do not display rubber-like structure even at higher temperatures. In such case, we talk of softening with a range of temperature. T_f of amorphous polymers increases with increasing molecular mass. These fact go a long way in the synthesis of polymer with desired properties. In the case of polydisperse polymers, polymer fractions of different molecular mass pass into visco-fluid state at different temperatures.

GLASS TRANSITION TEMPERATURE AND CHEMICAL CONSTITUTION OF POLYMERS

Polymers with polar molecules give easily glassy state because of high interaction energy. The links between polar groups of one chain with polar group of the other neighbouring chain are stronger and do not break unless the thermal motion is very intense. Thus, they form local cross-link points in the polymer which are not constant in time and go on breaking and forming a new. The average life time of groups in the linked state increases with the decrease in temperature and becomes appreciable near the glass transition temperature.

The formation of crosslinks limits their thermal motion. This results in making the entire system more rigid, even in presence of small number of crosslinks, forming a structure of fixed random arrangement of polymer molecules. The polymer acquires the properties of a solid below T_g. It has also been observed that in glassy polymers most of the hydroxyl groups takes part in hydrogen bonding. Branched-chain and linear aliphatic resins provide low crosslinks and higher chain flexibility. One group on a given molecule with more than one reactive site on the adjacent molecule tends towards cyclization process rather than crosslinking. The effect of the types of linkages and T in the case of epoxy resin.

FORCED ELASTICITY

Low molecular glasses like resin and silicate glasses do not respond to large stresses. Macromolecular glasses often becomes brittle at temperatures much below T_g. The ability of glassy polymers to deform considerably without breaking is responsible for their wide use in engineering. The large deformation of glassy polymers was at one time attributed to flow process called cold flow, however, mutual displacement of macromolecules is not probable. Methyl methacry late under a strain at temperature below T_g was found to recover its initial shape and size after heating above T_g.

This recoverable nature of large deformation of high molecular glasses suggests that the regularities of rubber-like state apply to the glass of state as well, showing that the glass transition is not a phase transition and a polymer possesses the same structure at temperatures above and below T_g. The relaxation time of polymer in glassy state is very great.

Equation 1 gives a relation between relaxation time and temperature

$$\tau = \tau_0 \, C^{\Delta u - \alpha s/kT}$$

where τ is the relaxation time, ΔU is the activation energy, a is the stress and a is a constant.

At small stresses, stress value does not affect relaxation time but at larger stresses τ becomes smaller and highly elastic strain appears. It can also be seen that after the removal of deforming stress, elastic strains do not disappear at temperatures below T_g.

STRESS STRAIN DEPENDENCE

Glassy polymers like polystyrene, poly (methyl methacrylate), poly (vinyl chloride), etc. on deformation begins to extend non-uniformly when a particular stress is reached. One part of specimen becomes thinner than the rest of the mass and a neck is formed. Glassy polymers like cellulose acetate, cellulose nitrate do not form neck upon forced elastic deformation. The stress at which the total rate of deformation becomes equal to force elastic deformation is called ultimate forced elasticity. The ultimate forced elasticity has been found to rise with increasing straining rate indicating its relaxation nature.

GASEOUS PHASE STATE

In 1654, Otto von Guericke performed his famous *Magdebur hemisphere* experiment to show that the air exerted pressure. The actual development of the modem the theories began with the work of E. Torrcelli (Italy) in 1643. Torricelli measured the atmospheric pressure with the help of a barometer.

GASES CHARACTERISTIC

All gases show some common characteristics. These are described below :

(i) Gases maintain neither the volume nor shape, and completely fill the container in which they are introduced.

(ii) Gases expand appreciably on heating.

(iii) Gases are highly compressible, that is, when the pressure is increased, the volume of a gas decreases.

(iv) Gases diffuse rapidly.

MEASURABLE PROPERTIES

The behaviour of gases can be described in terms of certain parameters. These are described below.

Mass and Amount

The mass of a gas can be easily determined by weighting the container containing the gas and then emptying the container by taking out the gas and weighting the empty container again the difference between the two weights give the mass of the gas.

In chemistry, according to the IUPAC recommendations, the amount of a substance is expressed in terms of the number of moles (n). The number of moles can be obtained from the mass of the gas-by using the relationship,

No. of moles (n)

$$= \frac{\text{Mass of the gas in gram (m)}}{\text{Molar mass of the gas in gram per mole (M)}}$$

Volume (V)

Volume of the as is equal to the volume of its equation. Volumes can be expressed in litre (L), millilitre (mL), or cubic centimetre (cm^3). The SI unit of volume is cubic metre (m^3). In common practice, however smaller units such as cubic decimetre (dm^3) and cubic centimetre (cm^3) are more frequently employed. These are related to each other as,

$$1 \text{ m}^3 = 10^3 \text{ dm}^3 = 10^6 \text{ cm}^3$$

The commonly used unit, litre (L) has now been redefined so that,

$$1 \text{ L} = 1 \text{ dm}^3$$

and 1 mL = 1 dm^3

The volume (V) of any gas depends upon its amount, temperature and pressure. Mathematically, volume of a gas is a function of the amount, temperature, and pressure, *i.e.*,

$$V = f\ (\text{amount, temperature, pressure})$$

or $$V = f\ (n, T, P)$$

The instrument used for the measurement of the pressure of a gas is called a manometer. It simply consists of a U-shaped tube containing mercury usually. One limb of the tube is longer than the other. Two types of manometers are used. These are:

(i) Those in which the longer limb is closed.

(ii) Those in which the longer limb is open.

Closed limb manometer is used only for gases at pressures less than the atmospheric pressure. The open limb manometer is used for all cases. In a case when the gas pressure is greater than atmospheric pressure, mercury stands at a higher level in the longer limb.

In such a case, the difference of levels is added to the atmospheric pressure to get the pressure of the gas. As pressure is force per unit area, the pressure obtained in terms of the height of the mercury column can be converted into force per unit area as follows:

Suppose height of the mercury column	= h cm
Area of cross-section of the tube	= Acm^2
∴ Volume of the mercury column	= $A \times hcm^3$
If density of mercury at room temp	= ρ g cm^{-3}
then mass of the mercury column	= $A \times h \times \rho$ gram

∴ weight of the mercury column = $(A \times h \times \rho) \times g$

where g if, the Gaseous State due to gravity.

This weight of the mercury column is the force acting on A cm^2. Hence.

$$\text{Pressure (P)} = \frac{\text{Force}}{\text{Area}} = \frac{A \times h \times \rho \times g}{A} = h\rho g$$

A *standard or normal atmospheric pressure* is defined as the pressure exerted by a mercury column of exactly 76 cm at 0°C. This is the pressure exerted by the atmosphere at the sea level.

The smaller unit commonly employed in expressing the pressures of gas is mm or torr (after the name of Torricelli, who invented the barometer). Hence,

$$1 \text{ atm} = 76 \text{ cm} = 760 \text{ mm or } 760 \text{ torr}$$

The SI unit of pressure is pascal (Pa) which is defined as the pressure exerted by a force of 1 newton on an area of 1 m^2.

The two units are related as

$$1 \text{ atm} = 101325 \text{ Pa or N m}^{-2}$$

$$= 1.01325 \times 10^5 \text{ Pa or Nm}^{-2}$$

For most of the purposes, the approximate value used is

$$1 \text{ atm} = 10^5 \text{ Pa or Nm}^{-2}$$

A bigger unit, called kilopascals (written as kPa) is also sometimes used. Thus

$$1 \text{ atm} = 10^2 \text{ kPa}$$

Measurement of Temperature : Temperature is a measure of the extent of hotness or coldness of a body. The measurement of temperature is based upon the principle that substances expand on heating. The most common substance whose expansion is made use of in the measurement of temperatures is 'mercury'. There are three different scales on which the temperatures are measured. These are

(i) Centigrade or Celsius scale (after the name of Anders Celsius)

(ii) Fahrenheit scale (after the name of Daniel Fahrenheit, a German instrument maker)

(iii) Kelvin scale (after the name of Lord Kelvin)

The Celsius scale is based upon taking the freezing point of water as 0°C and the boiling point of water as 100°C at normal atmospheric pressure (*i.e.* at sea level) and then dividing the range into 100 equal parts so that each part represents 1°C.

The Fahrenheit scale is based upon taking the freezing point of water as 32°F and the boiling point of water as 212°F and dividing the range into 180 equal parts. Thus, A large number of chemicals that go to form the basis of bigger molecules required for the synthesis of high polymers exist in gaseous state at the room temperature and atmospheric pressure, *e.g.*, acetylene, ethylene, propylene, 1-Butene, etc. In order to follow the mechanism of the formation of these bigger molecules and their subsequent polymerization, it is essential that we may not only study the characteristics and behaviour of these gases, but also the physical laws that govern their behaviour.

One of the main objectives of modern chemistry is to relate the properties of a large collection of molecules which constitute the matter in bulk (macroscopic properties) to the properties of individual atoms and molecules (microscopic properties). Inspite of the fact that the particles of a gas are always in random motion, a relation between microscopic and macroscopic behaviour is simpler in gases than in solids or liquids. The molecules of a gas, even at ordinary temperature and pressure, move with considerable kinetic energy, but they are least affected by the forces of attraction and repulsion.

The gases do not possess volume of their own but have attributes like touch and sound. They possess properties like mass, compressibility, refractive index, density and even electrical conductivity at very low pressure when the molecules get charged. Their behaviour has been found to depend on the population density of the molecules in a specified space.

TYPE OF GASES

There are two types of gases viz. *uniform* and *non-uniform*. A gas need not normally have its mass, pressure, volume and temperature the same at all the places. Generally, it is a mixture of several elements as air is a mixture of oxygen, nitrogen, water vapour, carbon dioxide, etc. Its composition vary from place to place. Even in a chemically pure substance, *e.g.*, nitrogen, it may be considered that either the mass or volume or temperature or pressure or all of them may vary from place to place. Such gases are called non-uniform gases. It is always possible to find out a small volume element where all the above measurable properties are constant. Such gas element are called uniform gases.

PROPERTIES OF GASES

There are two classes of properties viz. extensive and intensive.

The properties that depend on the amount of the gas like volume, mass, etc., are called extensive properties while in a uniform gas the properties that are independent of the amount of the gas, *e.g.*, pressure, temperature are called intensive properties. It will not be possible to define a uniform gas. It is one that has the value of each of its intensive property the same irrespective of the region of the gas.

STATE OF A GAS

Each condition of a gas, *i.e.*, mass, volume, temperature, pressure, etc., which is completely distinguishable from any other condition is called as the state of the gas. The above measurable properties are called the state variable of the gaseous system. The state of a system is a function of its state variables.

Mass

The gas is composed of innumerable molecules each having a mass of the order of 10^{-24}g. So all these molecules collectively contribute towards the mass of the gas.

Occupies Space

The space occupied by all molecules of the gas is very small as compared to the space occupied by the gas. It is roughly of the order of 1000th part of the space occupied by the gas. The remaining space at any instant is vacant, and as such it occupies a volume. In this context it would not be wrong to approximate the molecule as a point in the whole space. The volume is measured as millilitres.

Motion

At any temperature above absolute zero, the molecules of a gas are in motion. At room temperature the molecules of air move at a speed of 5×10^4 metres per second (1100 miles per hour), so they have fluidity.

Diffusivity

The process by which a gas moves from a high pressure through a porous wall is known as diffusion. The molecules of a gas are moving at a fast speed and do not occupy any appreciable volume by themselves, hence, they will not oppose expansion. So they will have very high diffusivity.

Isotropic

The molecules in a gas, occupying insignificant space, are moving randomly with high speed and as such do not have any effective attractive or repuslive force between each other with the difference that if one molecule moves in a stipulated direction then it must correlate with the other, moving simultaneously in the opposite direction. Thus, they will occupy the space of the container and as such will not have a volume of their own and can be easily added or separated if no change of composition is to be achieved. They are mutually soluble as there is too much vacant space in between them. So its movement appears to be equal in all the directions. Since they have the same movement in all directions, they are known as isotropic.

Stationary State

Any change in the motion of one molecule is correlated with the compensatory change in the other molecule. Therefore, unless external factor is introduced the state of the gas remains stationary.

Stationary Energy

The molecules of the gas collide with the walls of the container and rebound. These collisions are assumed to be perfectly elastic and as such exert a pressure on the walls of the container but do not lose energy on rebounding from it. Therefore, at a given temperature the total energy of the system remains constant.

LAWS OF NEWTON

The olecules of the gas change direction on striking against the wall of the container. In all these cases they are guided by the Newton's laws of motion. These laws provide equation of state of uniform gases at low pressures.

The behaviour of a uniform pure gas like hydrogen, oxygen, nitrogen, carbon dioxide, isopentane, etc., is governed by pressure and temperature as intensive variable and mass and volume as extensive variable. This can be mathematically expressed by the equation:

$$F(m, V, P, t) = 0 \quad \text{...(i)}$$

Since V and m have extensive nature

$$V = V(m, P, t) \quad \text{...(ii)}$$

$$= mV(1, P, t) \quad \text{...(iii)}$$

Eqn. (iii) represents a function depending on P and t as variables and refers to a unit mass which can be written as

$$V = m\overline{\upsilon}\ (P, t) \quad \text{...(iv)}$$

where $\overline{\upsilon}$ is the volume for a unit mass and is known as specific volume of the gas.

KINETIC MOLECULAR GAS MODEL

In the preceding sections of this chapter, we examined the physical behaviour of gases from an empirical point of view and paid no attention to questions like why gases occupy the entire space available to them, why they are highly compressible, how gases exert pressure, why different gases diffuse at different rates, etc. Since it is not possible to deduce the nature of gases from the measured properties alone, it becomes necessary to look for a model to account for the observed facts. In order that the model is simple, certain assumptions will have to be made about the nature of gases. We shall now examine such a model first by discussing the validity of assumptions and later by comparing its predictions with observed facts.

Assumptions of Kinetic Theory

The following assumptions are made to describe an ideal gas :

(i) A gas consists of a large number of particles (atoms or molecules) that are so small and so far apart (on the average) that the volume of the molecules is negligible compared to the empty space between them. This assumption seems to be justified in view of the great compressibility of the gas. Calculations also show that about 99.9% of the total volume in nitrogen gas at STP is empty space. The average spacing between molecules of nitrogen at STP is worked out to be about ten times the molecular diameter. When the gas is compressed, the average spacing between molecules is reduced.

(ii) There are no attractive forces between molecules and the molecules are completely independent of one other. The support for this assumption comes from the observation that even highly compressed gases expand spontaneously to occupy all the space available. This suggests that there is no appreciable interaction between molecules.

(iii) The molecules of a gas move randomly in straight lines colliding with one another and also with the walls of the container. The pressure exerted by a gas is the result of collisions of the molecules with the walls of the container. That particles of a gas are not at rest but in a state of continuous motion (called thermal agitation) cannot be directly observed because molecules are too small to be seen individually even through the best microscope. However, movement of particles of size in the neighbourhood of 1 p.

diameter (*e.g.*, particles of smoke floating in air) can be observed microscopically. This phenomenon is called Brownian motion (after the British botanist who first observed the irregular motion of pollen particles suspended in water). It is found that the smaller the particles, the more violent is the irregular motion; also, the higher the temperature the more vigorous is the movement of the particles. In the absence of attractive forces between molecules, it is reasonable to assume straight line random motion for the molecules which are much smaller in size. Molecules, however, change the path as a result of collision with other molecules or the walls of the container.

(iv) In these collisions, there is no net loss of kinetic energy, although, there may be a transfer of energy between the partners in the collisions (elastic collisions). If there were loss of kinetic energy in collisions (inelastic collisions), the motion of molecules would eventually stop and the pressure and temperature would drop to zero. Since the pressure of a confined gas does not change with time (unless we heat or cool it), we conclude that a gas exerts pressure through elastic collisions with the wall.

(v) At any particular instant, different molecules in a sample of gas have different speeds and hence, different kinetic energies, but the average kinetic energy of the molecules is directly proportional to the absolute temperature.

Root Mean Square Velocity

The molecules of a gas in an enclosed space suffer continuous changes of momentum through collision with other molecules. As a result of such collision any individual molecule will acquire velocities ranging from zero to infinity. It is quite probable that any molecule in a given time may exhibit an average velocity zero as there are equal probability of its movement in any direction (both positive and negative) but the average of its square velocities shall always be positive.

$$\sqrt{u^2} = \sqrt{\frac{u_1^2 + u_2^2 + u_3^2 + u_4^2 \ldots u_n^2}{n}} \qquad \ldots(i)$$

Distribution of Speed

The assumption that at any instant different molecules have different speeds and hence, different energies is valid in view of the great number of molecular collisions. As molecules collide with each other, we would expect the speed of the molecules to change. All molecules in a container will, therefore, not have the same speed at any one time. Even if the molecules had the same initial speed, the effects of collisions would soon disrupt such uniformity. The speed of each gaseous molecule probably changes billions of times during a second and we have no way of knowing the speed or kinetic energy of each individual molecule. Instead, we shall make a statistical approach to predict as to how many of the molecules have a particular speed and kinetic energy. We shall now examine the distribution of molecular speeds briefly.

In order to understand the distribution of molecular speeds, it may be helpful to look at a familiar distribution. As an example, we consider the distribution of performance of students in a class test. Let us suppose that there are 100 students and we have information about the marks (percentage-wise) obtained by each student.

It is difficult to obtain a reasonably accurate idea of how the marks are distributed among the students by looking at the marks of each student. For the purpose of obtaining a better idea of how the marks are distributed in the class we study the performance distribution of the students by grouping together all students with marks within a certain range. Let this be 10. Then all students having marks ranging between, say 10-20%, 20-30%. The same marks and the students in the class are 'distributed' over this range of marks. The height of each rectangle refers to the number of students having marks within the range indicated by the base of the rectangle.

Random Walk Model

Let the molecules be rigid spheres of diameter d moving at random at an average velocity $\overline{C}$ like Brownian motion. The two spheres will collide only whenever the distance between their centres will become d. Let the centre of a sphere be surrounded by an imaginary sphere of radius d. As the sphere is travelling with an average speed $\overline{C}$, in a unit time it will pass through a volume $\pi Nd^2\overline{C}$. Let there be N molecules per c.c. Then the sphere will experience $\pi Nd^2\overline{C}$ collisions per second. The quantity ndl is known as collision cross-section for the rigid sphere. It can be proved that the relative velocity after collision becomes $\sqrt{2\overline{C}}$.

The distance travelled by a sphere between collisions can be given by $\frac{1}{\sqrt{2}\ N\pi d^2}$ with velocity U_x, then this velocity U_x receives increments of velocity from random collision with other molecules. Feller has shown that the probability that a molecule will attain a velocity U_x can be given as:

$$p(U_x) = (2\ \pi\ \alpha^2)^{-1/2} \exp\left(U_x^2/2\ \sigma^2\right) \quad ...(i)$$

Where a is known as the standard deviation from the mean. We also know that in one dimension

$$\frac{1}{2} \text{Nm}\ \overline{V}^2 = \frac{1}{2} RT \quad ...(ii)$$

or

$$\overline{V}^2 = \frac{R\ T}{N\ m} \quad ...(ii)$$

Since velocity can be both positive or negative

$$\sigma^2 = V^2 \quad ...(iv)$$

Therefore, the probability that a molecule has a velocity between U_x and $U_x + dU_x$

$$P(U_x)\ dU_x = \left(\frac{m}{2\pi KT}\right)^{1/2} \exp\left(-mU_x^2/2\ KT\right) dU_x \quad ...(v)$$

If there are large number of molecules N, then the fraction

$$\frac{dN(x)}{N} = \left(\frac{m}{2\pi KT}\right)^{1/2} \exp\left(-mCUx^2/2\,KT\right) dU_x \quad ...(vi)$$

Jeans has applied similar arguments to two dimensions where the simultaneous velocity components are U_x and U_y, then the velocity C regardless of the direction is

$$C^2 = U_x^2 + U_y^2 \quad ...(vii)$$

$$\frac{DN(c)}{N} = \frac{m}{2\pi KT} \exp\left(-C^2/2KT\right) Cdc \quad ...(viii)$$

Extending the arguments to three dimensions

$$c^2 = U_x^2 + U_y^2 + U_z^2\ \frac{dN(C)}{dN} \quad ...(ix)$$

$$= 4\pi\left(\frac{m}{2\pi KT}\right)^{3/2} \exp\left(-mC^2/2KT\right)c^2 dc \quad ...(x)$$

gives the plot of equation (x) at different temperatures.

Most Probable Velocity

The value of C, for which $\frac{dN(C)}{dN}$ is maximum, is called the most probable velocity C_p. It is given by

$$\frac{-dC_p}{dC} = \frac{-m}{2\pi KT} \text{ [exp } (-mC^2p/2KT)\ 2C_p\ (C^2{}_p)$$

$$+ \text{ [exp } (-mC^2{}_p/2\ KT)]\ (2C_p) \qquad ...(i)$$

Since $\frac{-dC_p}{dC} = 0$

$$C_p^2 = \frac{2KT}{m} \qquad ...(ii)$$

Average Velocities

In this case of three-dimensional motion there are two types of average velocities viz. the average velocity C and average root mean square velocity $\left(\overline{C}\right)^{1/2}$.

$$\overline{C} = \frac{1}{N}\int_0^{\alpha} CNd(C) \qquad ...(i)$$

From equation (viii),

$$\overline{C} = 4\pi\left(\frac{m}{2\pi KT}\right)^{3/2}\int_0^{\alpha}\exp\left(-mC^2/2\ KT\right)C^3\,dC \qquad ...(ii)$$

Let $\frac{mC^2}{2KT} = x^2$

$$Cdc = 2KT\,\frac{x}{m}dx \qquad ...(iii)$$

or $C^2dc = \left(\frac{2KT}{m}\right)^2 x^3\,dx$

$$\overline{C} = 4\pi\left(\frac{m}{2\pi\,KT}\right)^{3/2}\left(\frac{2KT}{m}\right)^2\int_0^{\alpha}x^3\,\exp\left(-x^2\right)dx$$

since $x^3\,\exp\left(-x^2\right)dx = \frac{1}{2}$

$$\overline{C} = 2\pi\left(\frac{m}{2\pi\,KT}\cdot\frac{2KT}{m}\right)^{3/2}\left(\frac{2KT}{m}\right)^{1/2} = \left(\frac{8KT}{\pi m}\right)^{1/2} \qquad ...(iv)$$

$$\overline{C}^2 = 4\pi\left(\frac{m}{2\pi KT}\right)^{3/2}\int_0^{\alpha}\exp\left(-mC^2/2KT\right)C^4\,dc$$

$$\text{Let} = \frac{mC^2}{2KT}X^2$$

$$Cdc = \frac{mKT}{m}Xd_x$$

From equation (54)

$$C^3 = \left(\frac{2KT}{m}\right)^{3/2}X^3$$

$$C^4\,dC = \left(\frac{2KT}{m}\right)\left(\frac{2KT}{m}\right)^{3/2}X^4 \quad \text{...(v)}$$

$$\overline{C}^2\,4\pi = \left(\frac{m}{2\pi KT}\right)^{3/2}\left(\frac{2KT}{m}\right)^{5/2}\int_0^{\alpha}X^4\exp\left(-x^2\right)d_x \text{ ...(vi)}$$

But $X^4\exp\left(-X^2\right)d_x = \frac{3}{8}\sqrt{\pi}$

$$\overline{C}^4 = 4\pi\times\frac{3}{8}\sqrt{\pi}\left(\frac{m}{2\pi KT}\right)^{3/2}\left(\frac{2KT}{m}\right)^{5/2} = \frac{3KT}{m} \text{ ...(vii)}$$

$$= \frac{3PV}{Nm} = \frac{3P}{\rho} \quad \text{...(viii)}$$

where ρ is the density of the gas and is given by $\frac{Nm}{V}$

The three characteristic velocities of a gas in three dimensions are:

Most probable velocity $C_P = \left(\frac{2\,KT}{m}\right)^{1/2}$

Average velocity $\overline{C} = \left(\frac{8\,KT}{\pi m}\right)^{1/2} = 1.13\,C_p$ ...(xi)

Root mean square velocity $\left(\overline{C}\right)^{1/2} = \left(\frac{3\,KT}{m}\right)^{1/2} = 1.22\,C_p$

i.e., $\left(\overline{C}\right)^{1/2} : \overline{C} : C_p = 1 : 0.92 : 0.815$

The distribution of the three velocities and the effect of temperature on C are shown in Fig. 3.9b where $\frac{\Delta N}{N}$ is plotted against molecular speed.

Average Kinetic Energy

$$\overline{K}.\overline{E}. = \frac{1}{2}\,\overline{m}\overline{C}^2 = \frac{m}{2N_0}\int_0^{\alpha} C^2 dm$$

$$= \frac{3}{2}\,KT = \frac{3}{2}\left(\frac{R}{N_o}\right)T \qquad \text{...(x)}$$

The average kinetic energy per molecule of an ideal gas is proportional to the absolute temperature and is independent of the kind of molecule.

It can be seen from the above that the fraction of the molecules depends on two factors.

(1) The square of the velocity

(2) $\exp\left(-\frac{1}{2}mC^2/KT\right)$

We know that $\frac{1}{2}\,mC^2 = \frac{1}{2}\,mX^2 + \frac{1}{2}\,m^2 + \frac{1}{2}mZ^2$

In the random walk model one velocity component is independent of the value of the other. Therefore, the kinetic energy of a molecule is the sum of its kinetic energy components and the magnitude of one velocity component is independent of the value of other velocity components.

Similarly, $\exp\left(-\frac{1}{2}\,mC^2\;KT\right) = \left(\exp - \frac{1}{2}\,\frac{mX^2}{KT}\right)$

$$\left(\exp -\frac{1}{2}\,\frac{mY^2}{KT}\right)\left(\exp - \frac{1}{2}\,\frac{mZ^2}{KT}\right) \qquad \text{...(xi)}$$

The kinetic energies associated with the velocity components must be additive but the probabilities of their individual values must be multiplicative.

DERIVATION OF THE EQUATION OF STATE OF AN IDEAL GAS

Let there be a spherical vessel of radius r. The molecules of gas are colliding with the surface of the vessel with velocity $\overline{U}$. Any force exerted on the surface of the wall by a colliding particle is balanced by an external opposite force. The net force being then zero, no motion of wall occurs. Let us isolate a gas particle and follow it. The particle that elastically collides with the wall of the vessel rebounding at an angle ϕ, equal to its angle of approach. Since the sphere is symmetrical the particle will always be in the same plane passing through the centre of the sphere. The distance between collisions is 2r cos ϕ. If δt is the time between the collisions, then

$$\delta t = \frac{c \cos \phi}{\overline{U}} \qquad ...(i)$$

The number of collision per unit time is

$$\frac{1}{\delta t} = \frac{\overline{U}}{2n \cos\phi} \qquad ...(ii)$$

The component of the velocity along the direction perpendicular in the spherical surface is U cos ϕ. The component of the momentum of the particle of mass m before collision is –mU cos ϕ and + mU cos ϕ after the collision. The change in the momentum = 2mU cos ϕ

The force exerted by the particle on the wall

$$= \frac{2, U \cos \phi \times \overline{U}}{2r \cos \phi} = \frac{m\overline{U}^2}{r} \qquad ...(iii)$$

The outward force due to collision of A molecules

$$= \frac{A m\overline{U}^2}{r} \qquad ...(iv)$$

The area of the sphere = 4 π r^2

$$\text{Pressure} = \frac{Am\overline{U}^2}{4\pi r^2 \times r} = \frac{am\overline{U}^2}{4\pi r^3} \qquad ...(v)$$

The volume V of the sphere

$$= \frac{4}{3}\pi r^3$$

$$\text{Pressure P} = \frac{Am\ \overline{U}^2}{3V} \qquad ...(vi)$$

or $$PV = \frac{1}{3} A\ m\overline{U}^2 = \frac{2}{3} A \left(\frac{1}{2} m\overline{U}^2\right)$$

$$= \frac{2}{3}\ A.K.E.\ (\text{Kinetic energy}) \qquad ...(vii)$$

= RT for 1 mole of gas ...(viii)

Average kinetic energy $= \frac{3}{2}$ RT

or T is a measure of KE (Kinetic energy).

Since in a gas at a fixed temperature A, w and $\overline{U}$ are fixed PV is constant which verifies the Boyles law.

At constant pressure and temperature. V is proportional to T which verifies the Charles law.

At constant pressure and temperature, V is proportional to A, which verifies the Avogadro's law.

Collision Frequency

Let us consider the molecules hitting a plane surface.

As already explained that all directions in space are equivalent, the results will be expected independent of direction. According to Boltzmann distribution law the molecules with a volecity of w,

$$\frac{dN(u)}{N} = \left(\frac{m}{2\pi KT}\right)^{1/2}\left[\exp\left(\frac{-mu^2}{2KT}\right)\right] \quad ...(ix)$$

The number dn of molecules per unit volume reaching the wall with velocities between u and u + du

$$dz(u) = N\left(\frac{m}{2\pi KT}\right)^{1/2}\left[\exp\left(\frac{-mu^2}{2KT}\right)\right]udu \quad ...(x)$$

z will give the total surface collision frequency per unit area, which can be obtained by integrating the above from zero to infinity.

$$z = N\left(\frac{KT}{2\pi m}\right)^{1/2} \quad z = N\left(\frac{KT}{2\pi m}\right)^{1/2} \quad ...(xi)$$

$$= \frac{1}{4}N\overline{V} = N\left(\frac{RT}{2\pi M}\right)^{1/2} \quad ...(xii)$$

$$= P\left(\frac{1}{2\pi m\ KT}\right)^{1/2} \quad ...(xiii)$$

The mass collision frequency mz is given by

$$mz = \rho\left(\frac{KT}{2\pi m}\right)^{1/2} = \rho\left(\frac{RT}{2\pi M}\right)^{1/2} \qquad ...(xiv)$$

where ρ is the density of the gas.

Bimolecular Collision Frequency

Molecules undergo collision as they move through the gas. This will result in the change of direction and speed of the colloiding molecules. They will behave as hard spheres of radius r and undergo elastic collision so that the initial and final kinetic energies do not change.

Collision Diameter

The molecules are treated as rigid noninteracting spheres. A moving molecule will collide with other molecule whose centres come within a distance d_{12}. Thus, the molecular collision diameter d_{12} is the minimum distance of approach of two molecules colliding with zero initial kinetic energy. If r_1 and r_2 are the radii of the two molecules, then d_{12} is equal to the sum of r_1 and r_2.

The quantity πd_{12} is called scattering cross-section of the molecules and is represented by σ_{12}.

$$mz = \rho\left(\frac{KT}{2\pi m}\right)^{1/2} = \rho\left(\frac{RT}{2\pi M}\right)^{1/2} \qquad ...(i)$$

where μ is the viscosity coefficient.

Number of Collisions

The molecule A is approaching towards molecule B and shall collide with it whenever the distance between their centres become d_{12}. Let $\overline{U}$ be the average relative speed and n the number of molecules per ml. In unit time the sphere of influence of A will sweep $\pi n d^2{}_{12}\overline{U}$. The number of collision experienced by A will be $\pi n d^2{}_{12}\overline{U}$ per second.

The relative velocities of two molecules just before and after the collision or just grazing collisions. The average may be taken as collision at 90°. Therefore, the relative velocity will be $2^{1/2}\overline{U}$.

The number of collisions z_1 will be given by

$$z_1 = 2^{1/2}\,\pi\, n d_{12}^2 \overline{U} \qquad ...(i)$$

but $$= \overline{U}\left(\frac{8\,KT}{m}\right)^{1/2} \text{ and } n = \frac{P}{KT}$$

$$z_1 = 2^{1/2}\, d_{12}^2\, \pi\left(\frac{8\,KT}{m}\right)^{1/2} \quad n = \frac{P}{KT}$$

$$= 4\, d_{12}^2\, \pi\left(\frac{\pi}{mKT}\right)^{1/2} \qquad \text{...(ii)}$$

From Eq. (i)

$$= \frac{m}{2(2)^{1/2}\, d_{12}^2}\left(\frac{8\,KT}{m}\right)^{1/2} = \left(\frac{8KT}{\quad}\right)^{1/2} \frac{1}{d_{12}^2}$$

$$z_1 = 4P \qquad \text{...(iii)}$$

Similarly, when the two molecules are similar

$$Z_{11} = \frac{1}{2}(2)^{1/2}\, \pi n^2\, \overline{V} \qquad \text{...(iv)}$$

In case the two molecules are different, then the number of collisions

$$= 2(2)^{1/2}\, \pi d_{12}^2 \overline{U}\, n_1 n_2$$

$$Z_{12} = 2(2)^{1/2}\, \pi d_{12}^2 \left(\frac{8\,KT}{\pi\, mr}\right) n_1 n_2$$

$$= \left(\frac{n_1}{V}\right)\left(\frac{n_2}{V}\right) 4\,\pi\, d_{12}^2 \left(\frac{KT}{2\pi\, m_r}\right)^{1/2} \qquad \text{...(v)}$$

where reduced mass $m_r = \dfrac{m_1\, m_2}{m_1 + m_2}$ and $d_{12} = \dfrac{d_1\, d_2}{2}$

$$Z_{12} = \frac{4\,\pi\, d_{12}^2\, p_1 p_2}{KT(2\pi\, m_r\, KT)^{1/2}} \qquad \text{...(vi)}$$

where p_1 and p_2 are the respective pressures.

Mean Time Interval

The mean time interval between collisions

$$\tau = \frac{1}{Z} = \frac{(\pi\, m\, KT)^{1/2}}{4\,\pi\, d_{12}^2\, P} \qquad \text{...(vii)}$$

Mean Free Path

The mean free path between collisions

$$\lambda = \frac{\overline{U}}{Z_1} = \frac{1}{2^{1/2}\,\pi n d_{12}^2} = \frac{\sqrt{2}\,KT}{2\pi d_{12}^2\,P} \qquad \text{...(viii)}$$

The free path depends upon the number, density and size of the molecules.

If the molal volume of the gas is 50 ml/g mole then the diameter of a spherical molecule will be 5×10^{-8} cm. At NTP n is 2.7×10^{19} molecules/cm^3 and is 3×10^{-6} cm. A molecule of hydrogen will experience 10^{11} collisions/sec and the ternary collision frequency will be 10^{33} collisions per sec. cm^3.

SOLVED EXAMPLES

Example 1:

Equal volumes of an organic liquid and water gave 55 drops and 35 drops respectively. The densities of water and the organic liquid are 0.996 and 0.80g cm^{-3} and the surface tension of water is 7.2×10^{-2} Nm^{-1}. Calculate the surface tension of the organic liquid. How many times a water drop is heavier than a drop of the organic liquid ?

Solution:

$$\frac{\gamma_1}{\gamma_2} = \frac{n_2 d_1}{n_1 d_2}$$

or
$$\gamma_1 = \frac{\gamma_1 n_2 d_1}{n_1 d_1}$$

Since

$n_1 = 55$ $\qquad$ $n_2 = 35$

$d_1 = 0.80\text{g cm}^{-3}$ $\qquad$ $d_2 = 0.996\text{g cm}^{-3}$

$= 8.0 \times 10^2 \text{ kg m}^{-3}$ $\qquad$ $= 9.96 \times 10^2 \text{ kg m}^{-3}$

$\gamma_1 = ?$ $\qquad$ $\gamma_2 = 7.2 \times 10^{-2} \text{ Nm}^{-1}$

Therefore

$$\gamma_1 = \frac{(7.2 \times 10^{-2}\text{ Nm}^{-1})(35)\,(8.0 \times 10^2\text{ kg m}^{-3})}{(55)\,(9.96 \times \text{kg m}^{-3})}$$

$= 2.63 \times 10^{-2} \text{ Nm}^{-1}$

Now,
$$\frac{m_{\text{water}}}{m_{\text{organic liquid}}} = \frac{\gamma_{\text{water}}}{\gamma_{\text{organic liquid}}}$$

$$= \frac{7.2 \times 10^{2}\ \text{Nm}^{-1}}{2.63 \times 10^{-2}\ \text{Nm}^{-1}} = 2.74.$$

Example 2:

Two immiscible liquids, water and chlorobenzene are boiled at 97.89 kNm^{-2} pressure and the mixture boils at 363K. The ratio of the weight of chlorobenzene to water collected in the distillate is 2.47. Calculate the molecular weight of chlorobenzene if vapour pressure of water at this temperature is 70.11 kNm^{-2}.

Solution:

Vapour pressure of chlorobenzene

$$= 97.89\ \text{kNm}^{-2} - 70.11\ \text{kNm}^{-2} = 27.78\ \text{kNm}^{-2}$$

$$\frac{\text{Weight of chlorobenzene}}{\text{Weight of water}} = \frac{P^{\circ}_{A}\ M_{A}}{P^{\circ}_{B}\ M_{B}}$$

i.e.

$$2.47 = \frac{27.78\ \text{kNm}^{-2} \times M_{A}}{70.11\ \text{kNm}^{-2} \times 0.018\ \text{kgmol}^{-1}}$$

$\therefore$ M_A *i.e.*, molecular weight of chlorobenzene

$$= \frac{2.47 \times 70.11\ \text{kNm}^{-2} \times 0.018\ \text{kgmol}^{-1}}{27.78\ \text{kNm}^{-2}} = 0.1122\ \text{kgmol}^{-1}$$

$$\text{Relative molar mass} = \frac{0.1122\ \text{kgmol}^{-1}}{0.001\ \text{kgmol}^{-1}} = 112.2.$$

Example 3:

The radius of a given capillary is 1.05 × 10^{-4} m. A liquid whose density is 0.80g/cm^{3} rises in the capillary up to a height of 6.25 × 10^{-2}m. Calculate the surface tension of the liquid assuming the contact angle θ to be zero.

Solution:

Here $r = 1.05 \times 10^{-4}$m $d = 0.80$g/cm^3

$h = 6.25 \times 10^{-2}$m $= 8.0 \times 10^{2}$kg/m^3

$g = 9.8$ms^{-2}

Since, $\gamma = 1/2\ rhdg$

$$\therefore \quad \gamma = \frac{1}{2}(1.05\times10^{-4}\text{m})\ (6.25\times10^{-2}\text{m})\ (8.0\times10^{2}\text{kg m}^{-3})(9.8\text{ms}^{-2})$$

$$= 2.5725 \times 10^{-2}\text{kg s}^{-2}$$

$$= 2.5725 \times 10^{-2}\text{Nm}^{-1} \qquad (1\text{ N} = 1\text{kg ms}^{-2})$$

Example 4:

The surface tension of toluene at 293K is 0.0284 Nm $^{-1}$ and its density at this temperature is 0.866g cm^{-3}. What is the largest radius of the capillary that will permit the liquid to rise 2 × 10^{-2} m ?

Solution:

Since $\gamma = \frac{1}{2}$ rhdg

Therefore $r = \frac{2r}{hdg}$

Here $\gamma = 2 \times 10^{-2}$ m $\qquad$ h = 0.866 g cm^{-3}

$\gamma = ?$ $\qquad$ $= 8.66 \times 10^{2}$ kg m^{-3}

$\gamma = 0.0284$ N m^{-1} $\qquad$ g = 9.8 m s^{-2}

Substituting these values in the above equation, we get

$$r = \frac{2(0.084\ \text{N m}^{-1})}{(2\times10^{-2}\text{ m})\ (8.66\times1^{2}\ \text{kg m}^{-3})\ (9.8\ \text{ms}^{-2})}$$

$$= 3.347 \times 10^{-4}\ \text{kg}^{-1}\ \text{s}^{2}\text{m}^{2}$$

$$= 3.347 \times 10^{-4}\text{m} \qquad (1\text{N} = 1\text{ kg ms}^{-2})$$

$$= 0.03347\text{ cm}$$

Example 5:

At a pressure of 101.3kNm^{-2} a mixture of water and nitrobenzene distils over at 372K. The vapour pressure of water at 372K is 97.70 kNm^{-2}. Estimate the proportion by weight of nitrobenzene in the distillate.

Solution:

$$\frac{\text{Weight of nitrobenzene } W_A}{\text{Weight of water } W_B} = \frac{P^\circ_A\ M_A}{P^\circ_B\ M_B}$$

Molecular weight of nitrobenzene = 0.123 kgmol^{-1}

Molecular weight of water = 0.018kg mol^{-1}

$P^\circ_B = P - P^\circ_A = 101.3 \text{ k Nm}^{-2} - 97.70 \text{ kNm}^{-2} = 3.60 \text{ kNm}^{-2}$

$$\therefore \quad \frac{W_A}{W_B} = \frac{3.60 \text{ kNm}^{-2}}{97.70 \text{ kNm}^{-2}} \times \frac{0.123 \text{ gmol}^{-1}}{0.018 \text{ kgmol}^{-1}}$$

$\therefore$ WA : WB = 0.25 : 1.00

$\therefore$ Proportion of nitrobenzene in the distillate

$$= \frac{0.25}{1.25} \times 100 = 20\%.$$

Example 6:

(a) What is the average volume available to a molecule in a sample of nitrogen gas at STP?

(b) Assuming that N_2 molecule is approximately spherical with an effective diameter of 3.6Å, calculate

(i) The actual volume occupied by a N_2 molecule;

(ii) the percentage of the molar volume that is empty space;

(iii) The average distance between neighbouring gas molecules at STP.

Solution:

We know that one mole of an ideal gas at STP (*i.e.*, at 0°C and 1 atm pressure) occupies 22.4 litres. Therefore, assuming ideal behaviour for nitrogen, one mole of this gas will occupy 22.4 litres or 22,400 cm^3 at 0°C and 1 atm, pressure. There are 6.02×10^{23} molecules in one mole. Therefore, the number of molecules in 1 cubic centimeter is

$$\frac{6.02 \times 10^{23}}{22,400} = 2.69 \times 10^{19} \text{ molecules per cm}^3$$

and the volume available to a gas molecule is

$$\frac{1}{2.69 \times 10^{19}} = 3.72 \times 10^{-20} \text{ cm}^3 \text{ molecule}^{-1}$$

(b) Volume of a sphere $= \frac{4}{3} \pi r^3$

As diameter for nitrogen is 3.6 Å = 36×10^{-8} cm

r = radius of nitrogen molecule $= \frac{3.6 \times 10^{-8}}{2}$ cm $= 1.8 \times 10^{-8}$ cm.

Volume of a nitrogen molecule = $4/3\ (1.8 \times 10^{-8})^3$

$= 2.44 \times 10^{-23}$ cm^3 molecule^{-1}

Volume of 1 mole of nitrogen molecules

$= 2.44 \times 10^{-23} \times 6.02 \times 1023 = 14.7$ cm^3 mole^{-1}

We find that the actual volume occupied by 1 mole of molecules is 14.7 cm^3 while the gas molar volume is 22,400 cm^3. Thus, the difference 22,400 – 14.7 = 223,85.3 cm^3 is apparently the empty space.

$$\text{per cent empty space} = \frac{\text{Volume empty}}{\text{Volume available}} \times 100$$

$$= \frac{223,85.3}{224.00} \times 100 = 99.9\%$$

The average distance between any molecule and its nearest neighbour can be obtained from the volume available to a gas molecule: Let us assume that the volume available to a molecule has the shape of sphere of radius R. Hence, the volume available to a molecule is $4/3\ R^3$ and this is equal to 3.72×10^{-20} cm^3 as found in (a)

$$4/3\ R^3 = 3.72 \times 10^{-20}\ \text{cm}^3$$

$$R^3 = \frac{3 \times 3.72 \times 10^{-20}}{4x}$$

$$R = 20.7 \times 10^{-8}\ \text{cm}$$

Thus, on an average, the volume available to a gas molecule can be assumed to be the volume of a sphere with R = 20.7×10^{-8} cm^3, and accordingly the average distance between the centres of any two neighbouring molecules in the gas sample will be 2R. Therefore, the average distance between a gas molecule and its nearest neighbour is 2R = $20.7 \times 10^{-8} = 41.4 \times 10^{-8}$cm = 41.4Å.

Another way of computing the distance between a gas molecule and its nearest neighbour in a sample of gas is by assuming that the gas volume is divided into small cubes of edge-length and the volume of this small cube is the volume, on an average, available to a gas molecule.

The length of this small cube (1) will then roughly be equal to the average distance between the gas molecules assuming that the molecules are for most part evenly spaced at the centres of identical cubes. If 1^3 is the volume of each small cube, then the value of 1 can be computed from the relation

$$l^3 = 3.72 \times 10^{-20} cm^3$$

where 3.72×10^{-20} is the volume available to each molecule

or $l = 33.4 \times 10^{-8}$ cm $= 33.4$Å

From these calculations, we find that the average distance between two neighbouring molecules at STP is about 33.4×10^{-8} cm or 33.4Å (students should find out why this value is different from that obtained previously). The diameter of the gas molecule is only 3.6Å and the distance between molecules in the gas is, therefore, about ten times their diameter. Similar calculations for solid (density 1.03 gm/cm^3) and liquid (density 0.8 gm/cm^3) nitrogen show that in these states, the average distance between molecules is about 3.56Å and 3.8Å respectively (which are nearly equal to the molecule diameter of nitrogen). Hence, in solid or liquid states, individual molecules touch each other and it is difficult to compress the substances in these states unlike in the gaseous state.

Example 7:

At 20°C, water formed 29 drops when flowing through the capillary of a stalagmometer, while an equal volume of ether formed 86 drops. If the densities of water and ether are 0.997 and 0.70 gm per cc respectively, find the surface tension of ether if that of water is 72.8 dynes/cm.

Solution:

We have $$\gamma_2 = \frac{n_1}{n_2} \cdot \frac{\rho_2}{\rho_1} \cdot \gamma_1$$

$$= \frac{29}{86} \times \frac{0.07}{0.997} \times 718 = 17.24 \text{ dynes/cm}$$

Example 8:

At 293 K, $10^{-2} dm^3$ of water formed 29 drop sand the same volume of an organic liquid formed 86 drops in the same stalagmometer. Density of the organic liquid is 0.7 g cm^{-3}. If the surface tension of water is 7.2 $\times 10^{-2}$ N m^{-1}, what would be the surface tension of the organic liquid ?

Solution:

If n_1, ρ, and γ_1 are the number of drops, density and the surface tension of the organic liquid, and n_2, d_2 and γ_2 are the number of drops, density and the surface tension of water respectively, then we have

$$\frac{\gamma_1}{\gamma_2} = \frac{n_1 \rho_1}{n_1 \rho_2}$$

or $$\gamma_1 = \frac{\gamma_2 n_2 \rho_1}{n_1 \rho_2}$$

Since $n_2 = 29$ $n_1 = 86$

$\rho_2 = 1 \text{ g cm}^{-3}$ $\rho_1 = 0.70 \text{ g cm}^{-3}$

$= 1 \times 10^3 \text{gm}^{-3}$ $= 7.0 \times 10^2 \text{ gm}^{-3}$

$\gamma_2 = 7.2 \times 100^{-2}\text{Nm}^{-1}$ $\gamma_1 = ?$

Therefore

$$\gamma_1 = \frac{29(7.0 \times 10^2 \text{ kg m}^{-3})(7.2 \times 10^{-2} \text{ N m}^{-1})}{86\,(10^3 \text{ kg m}^{-3})}$$

$$= 1.726 \times 10^{-2} \text{ m}^{-1}$$

or $$= 17.26 \text{ dynes cm}^{-1}$$

Example 9:

A steel ball of density 7.90 gm per cc and 4 mm diameter requires 55 sees to fall a distance of 1 meter through a liquid of density 1.10 gm per cc. Calculate the viscosity of the liquid in poises.

Solution :

According to equation (8), we have

$$\eta = \frac{2r^2\,(\rho - \rho')g}{9v}$$

where, r = Radius = 0.2 cm

v = Velocity = 100/55 cm sec^{-1}

$$\eta = \frac{2 \times (0.2)^2\,(7.90 - 1.10) \times 980}{9 \times \frac{100}{55}} = 32.58 \text{ poises}$$

Example 10:

A sphere of 5 × 10^{-2} cm and density 1.10 gm per cc falls at constant velocity through a liquid of density 1.00 gm per cc and viscosity 1.00 poise. What is the velocity of the falling sphere?

Solution:

Rearranging equation (8) we have

$$v = \frac{2r^2\,(\rho - \rho')g}{9\eta}$$

On substituting

$$v = \frac{2 \times (5 \times 10^{-2})^2 \times (1.10 - 1.00) \times 980}{9 \times 1.00}$$

$$= 5.434 \times 10^{-3} \text{ cm/sec.}$$

Example 11:

Naphthalene on oxidation gives phthalic anhydride, a very important raw material for glyptal type of resins. When it is heated to 30°C, it exerts a vapour pressure of 0.177 mm. Calculate the effusion rate when M = 128.16 g/g.mole.

Solution:

$$0.177 \text{ mm} = 2.36 \times 10^5 \text{ g-cm/sec}^2$$

$$Z_m = \frac{\alpha}{4} P \left(\frac{3M}{RT}\right)^{1/2}$$

$$= \frac{0.92}{4} \times 2.36 \times 10^5 \left(\frac{3 \times 128.16}{8.314 \times 303 \times 10^7}\right)^{1/2}$$

$$= 0.23 \times 2.36 \times 10 \left(\frac{128.16}{8.314 \times 10^7}\right)^{1/2}$$

$$= 6.73 \text{ g/cm}^2 \text{ sec.}$$

Example 12:

In a steam distillation of an insoluble oil, the mixture boils at 368K when the external pressure is 100 kNm^{-2}. If the vapour pressure of water at this temperature is 84.0 kNm^{-2}. If the molecular weights of water and oil are 0.018 kg mol^{-1} and 0.380 kg mol^{-1} respectively, calculate the amount of steam required to distil 0.1 kg of oil.

Solution:

$$\frac{\text{Weight of steam } W_A}{\text{Weight of oil } W_B} = \frac{P^\circ_A \; M_A}{P^\circ_B \; M_B}$$

$$P^\circ_A = 84.0 \text{ kNm}^{-2}$$

$$P^\circ_B = 100\text{–}84.0 = 16 \text{ kNm}^{-2}$$

$$\frac{W_A}{W_B} = \frac{84.0 \text{ kNm}^{-2}}{16.0 \text{ kNm}^{-2}} \times \frac{0.018 \text{ kg mol}^{-1}}{0.380 \text{ kgmol}^{-1}}$$

Weight of oil in the distillate = 0.1 kg.

$$\therefore \quad W_A = \frac{84.0 \text{ kNm}^{-2}}{16.0 \text{ kNm}^{-2}} \times \frac{0.018 \text{ kg mol}^{-1}}{0.380 \text{ kgmol}^{-1}} \times 0.1 \text{ kg}$$

$$= 0.0248 \text{ kg}$$

i.e., weight of steam necessary to distil 0.1 kg of oil is 0.0248 kg.

Example 13:

Compute the relative rates of diffusion of hydrogen and carbon dioxide. Molecular mass of hydrogen is 2 and that of carbondioxide is 44.

Solution:

$$\frac{r_{H_2}}{r_{CO_2}} = \frac{\left(M_{CO_2}\right)^{1/2}}{\left(M_{H_2}\right)^{1/2}} = \left(\frac{44}{4}\right)^{1/2}$$

$$= \frac{6.63}{1.41} = 4.7$$

Hydrogen diffuses 3.7 times faster than carbon dioxide.

Example 14:

The speeds of diffusion of CO_3 and O_3 were found to be as 0.29 and 0.271. What is the molecular mass of O_3 if the molecular mass of CO_2 is 44?

Solution:

$$\frac{r_{O_3}}{r_{CO_2}} = \left(\frac{M_{CO_2}}{M_{O_3}}\right)^{1/2} \quad \frac{0.271}{0.29} = \left(\frac{44}{M_{O_3}}\right)^{1/2}$$

Squaring both sides

$$\frac{(0.271)^2}{(0.29)^2} = \frac{44}{M_{O_3}}$$

or $$M_{O_3} = 44 \times \frac{(0.29)^2}{(0.271)^2} = \frac{44 \times 0.29 \times 0.29}{0.271 \times 0.271}$$

$$= 50.4 \text{ (Mol. mass of ozone is 50.4).}$$

Example 15:

A sample of 1.0 g of gas contained in a 1.0 litre vessel at 24°C exerts a pressure of 0.836 atm. What would be the pressure of the sample if it was compressed to a volume of 0.9631 and heated to a temperature of 47°C.

Solution:

For the gas in the two states we have

Initial state: $P_1 = 0.836$ atm; $V_1 = 1.01$;

$n_1 = N$; $T_1 = 273 + 24 = 297°K$ $P_1V_1 = nRT_1$

Final state: $P_2 = ?$; $V_2 = 0.9631$; $n_2 = n$;

$T_2 = 273 + 47 = 320°K$ $P_2V_2 = nRT_2$

In this problem n is the same in the two states, therefore,

$$\frac{P_1V_1}{T_1} = \frac{P_2V_2}{T_2}$$

or

$$P_2 = \frac{V_1}{V_2} \times \frac{T_2}{T_1} \times P_1$$

Substituting the numerical values

$$P_2 = \frac{1.01}{0.9631} \times \frac{320°K}{320°K} \times 0.863 \text{ atm}$$

$$= 0.935 \text{ atm}$$

2

Phase Systems and Reactions

ONE-COMPONENT SYSTEMS

One component is the smallest number by means of which the composition of each phase is expressed, it is known as a one component system.

Maximum Number of Phases : When F is minimum, P becomes maximum. The minimum number of degrees of freedom possible in a system is zero. Consider an equilibrium state of a one component system having zero degree of freedom. According to the phase rule,

we have $$P = C - F + 2 \qquad ...(1)$$

or $$P = 1 - 0 + 2 = 3 \qquad ...(2)$$

Thus, the *maximum number of phases in any equilibrium state of one component system having zero degree of freedom will be three.* It cannot be more then three.

Maximum Number of F : The maximum value of F is expected in the case when P is minimum. As the minimum number of phases in any system is one, it follows from phase rule equation that the maximum degree of freedom in one component system are two.

$$F = C - P + 2 = 1 - 1 + 2 = 2.$$

TWO COMPONENT SYSTEMS

When two components denote the smallest number by means of which composition of each phase can be expressed, it is known as a two component system.

Maximum number of Phases : When F is minimum, P becomes maximum. The minimum number of degrees of freedom possible in any system is zero. Consider an equilibrium state of a two component system having zero degree of freedom. According to the phase rule, we have

$$F = C - P + 2 \quad ...(1)$$

$$0 = 2 - P + 2 \text{ or } P = 4$$

Thus, the maximum number of phases in any equilibrium state of a two component system having zero degree of freedom will be four.

Maximum number of F : The maximum value of F is expected in the case when P is minimum.

As the minimum number of phases in any system is one, it follows from equation (1) that maximum degree of freedom in a two component system are three.

$$F = C - P + 2 = 2 - 1 + 2 \text{ or } F = 3$$

Phase Diagram, As the value of F is three, it means that a phase is to be defined by three variables, *i.e.,*. pressure, temperature and composition. Pressure, temperature and composition could be represented by three co-ordinate axes at right angles, yielding models for two component systems. But the space models cannot be conveniently represented on paper.

In order to have simple phase diagram for two component systems, it is a usual practice to choose any two of three variables for graphic representation, assuming third to be constant. In this manner, we can have

(i) Pressure-temperature diagram (P–T)

(ii) Temperature-composition diagram (T–C)

(iii) Composition-pressure diagram (C–P).

In most of the studies, it is the usual practice to keep the pressure constant and thus, an T–C diagram is the one which is most often met.

Reduced phase rule : In all the above P–T. T–C and C–P diagrams, one of the variables is fixed. Thus the degree of freedom of the system is reduced by one. Therefore the phase rule equation (1) for the component system is written as

$$F - 1 = C - P + 1$$

$$F = C - P + 1 \quad ...(2)$$

where F' gives the remaining degrees of freedom of the system. Equation (2) is known as *reduced phase rule.*

TYPES OF TWO COMPONENT SYSTEMS

We will only discuss the solid-liquid systems of two components.

Potassium Iodide-water System

The potassium iodide-water system is a typical example of a two-component system which forms a eutectic mixture. Hence its phase diagram is similar to the Fig. 2.1. The various features of this system are illustrated in the T–C diagram, pressure being considered as one *atm* pressure. In the above diagram, the melting point of potassium iodide is not shown because it is very high.

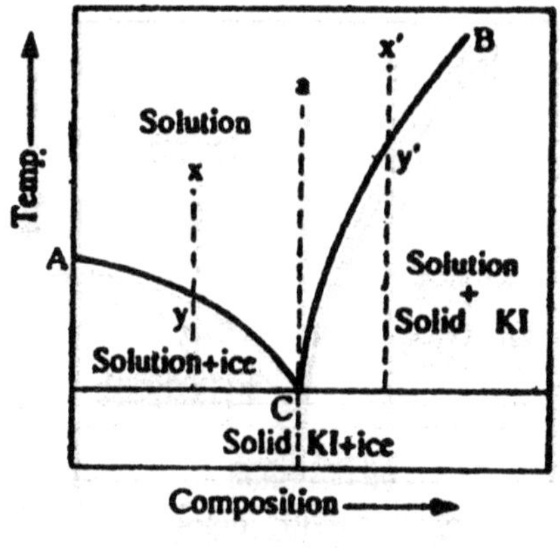

Fig. 2.1 : KI-H_2O System.

Curve AC : Point A represents the freezing point of water at 1 atm pressure. This temperature is 0°C.

When increasing quantities of potassium iodide are added, the freezing point of water is lowered along the curve AC. Thus, AC may be called the freezing point curve of water.

Along the curve AC, the solution of KI in H_2O is in contact with ice. Thus, there are two phases. Applying the condensed phase rule to any point on this area.

$$F' = C - P + 1 = 2 - 2 + 1$$

or $$F' = 1.$$

Thus, the system is univariant, *i.e.,* has one degree of freedom. It means that there will be a definite composition of the solution corresponding to each temperature. When further quantities of potassium iodide are added, the lowering of freezing point continues along the curve AC till the point C is reached. Here the solution becomes saturated with respect to solid potassium iodide and freezes at a constant temperature (–23°C) and fixed composition (52% KI).

Curve BC : This is known as the solubility curve of potassium iodide in water at different temperatures. The solubility increases with rise of temperature and becomes maximum at the point B which is the boiling point of the saturated solution. Allow a saturated solution of potassium iodide at B to cool. The solution will freeze along the curve BC. During this, solid KI crystallises out.

This crystallisation of KI continues till the temperature and composition corresponding to the point C is reached. At this point, the solution freezes as a whole with fixed composition. Along the curve BC, solid KI is in equilibrium with solution.

Again, applying the condensed phase rule to this system, we obtain

$$F' = C - P + 1 = 2 - 2 + 1 \text{ or } F' = 1$$

Thus the system is univariant, *i.e.,* has no degree of freedom. The steep rise of curve BC indicates that the solubility of KI increases slowly with the rise of temperature.

Point C : It is the eutectic or cryohydric point. At this point the two curves AC and BC intersect. This point corresponds to a definite temperature (–23°C) and definite composition (527% KI + 487% ice).

At this point C, ice, solution and solid KI are in equilibrium. This point has no degree of freedom which is evident from phase rule equation.

$$F' = C - P + 1 = 2 - 3 + 1 \text{ or } F' = 0$$

Thus, C is a non-variant point.

Area above AOB : If the point lies above the curve BC. It means that the dilute solution of KI is there. But the area above the curve BC represents the concentrated solution of potassium iodide.

Cooling : Three cases may arise :

(i) Consider a dilute solution represented by the point x. On cooling, the temperature will lower along the curve xy without any change in composition.

This continues till the point y is reached. At this point ice separates out. Here the system will become univariant and further cooling will take place along the curve yC. This continued till the point C is reached. At this point ice and K.I freeze as a whole with fixed composition, *i.e.,* eutectic mixture.

(ii) Consider a concentrated solution represented by the point x'. Allow it to cool. This cooling continues without any change in composition along the curve x'y'. At the point y', solid KI separates out.

Further cooling will proceed along the curve y'C until the point C is reached. At this point, ice and K[freeze as a whole to give the eutectic mixture [KI = 52% and ice = 48%].

(iii) Consider a solution which is represented by the point a. This point lies vertically above the eutectic point C. When the solution is allowed to cool, the temperature will lower along the curve aC until the eutectic point C is reached. At this point, ice and potassium iodide separate out simultaneously.

This mixture of potassium iodide and ice deposited at the eutectic point is known as *cryohydrate.*

Type IInd A.

Solid-liquid equilibrium of type IInd A :

A compound is said to hive a congruent melting point when it melts sharply at a constant temperature into a liquid of the same composition as that of solid from which it is derived.

Suppose A and B are two components and AB is the compound formed by their chemical combination. The compound AB has a sharp congruent melting point. The phase diagram for such a phase system is shown in Fig. (2.2).

Curve AC : Point A represents the melting point of pure component A. By gradual addition of B to A, the freezing point of A is lowered along the curve AC. Thus, the curve AC represents the freezing point curve of the component A. Along this curve, the solid A is in equilibrium with the solution of the component B in A. Therefore, the number of phases along AC is two. Applying the condensed phase rule to the system, we obtain

$$F' = C - P + 1 = 2 - 2 + 1 = 1$$

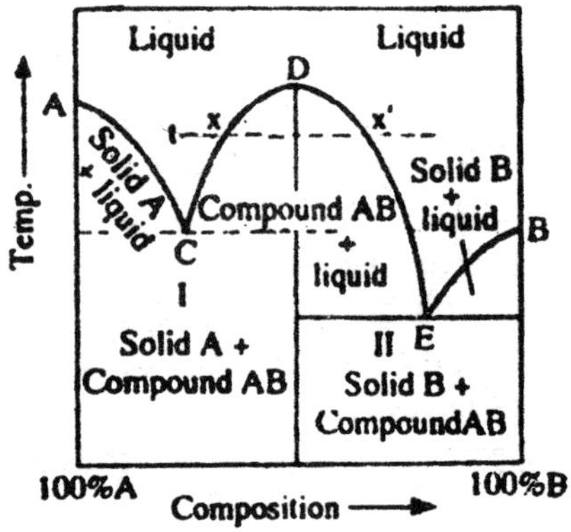

Fig. 2.2 : Solid-liquid equilibrium (Type II)

Thus, the system is univariant, *i.e.,* only the composition varies along AB.

Curve BE : Point B represents the melting point of pure component B. When A is added to B, the freezing point B is lowered along the curve BE. The curve is known as the *freezing point curve* of the component B.

Along this curve, the solid B is in equilibrium with the solution of the component A in B. Thus, the number of phases along BE is two. Applying the condensed phase rule to this system,

$$F' = C - P + 1 = 2 - 2 + 1$$

or $$F' = 1.$$

Thus, the system is univariant, *i.e.,* only the composition varies along BE.

Curve CDE : This curve is known as *freezing point carve* of the solid compound AB. Along this curve, the solid compound AB is in equilibrium with the liquid phase at different temperatures. The point D is known as *congruent melting* point of the compound because both the liquid and solid phases have the same composition at the point D. It means that two-component system becomes one component at this temperature as both the solid and liquid phases contain the same compound AB. Applying the phase rule to this point D, we get

$$F' = C - P + 1 = 1 - 2 + 1 = 0$$

Thus, the point D is non-variant : It means that this point D represents a definite melting point A.

Eutectic point C : The two curves AC and CDE intersect at the point C. At this point, solid A and compound AB are in equilibrium with liquid phase. Therefore, the number of phases is three. Now, applying the phase rule equation to this point C, we obtain

$$F' - C - P + 1 = 2 - 3 + 1 = 0$$

Thus, the system at C is invariant, *i.e.,* has no degree of freedom. It means that solid A, solid compound AB and liquid can only exist at a definite temperature and the composition of the liquid phase must also be definite.

Eutectic point E : The two-curves BE and CDE intersect at the point E where solid B, solid compound AB are in equilibrium with liquid phase. Again, the point E is invariant, *i.e.,* has no degree of freedom. The horizontal line passing through the eutectic point C and E is known as solid because only phase exists along it.

Areas I and II : The names of the various phases in areas I and II are shown in the Fig. 3.14.

Solid-liquid Systems of Two Components

Condensed systems : The effect of pressure on the equilibrium between liquids and solids is very small. Therefore, the effect of pressure on such systems may be disregarded. Therefore, the experiments in liquid-solid equilibria are usually conducted under constant atmospheric pressure. *"Such a system in which only solid and liquid phase are considered and the experiments are carried out under atmospheric pressure, is called a condensed system."*

The graphical representation of condensed systems of two-components is simplified as there are only two variables, *i.e.*, temperature and composition, keeping pressure to be constant and these can be represented on a plane paper in the case of a condensed system, variables gets fixed. In other words, one of the degrees of freedom of the system is reduced by one. Therefore, the equation (1) becomes as:

$$F - 1 = C - P + 1$$

or $$F = C - P + 1 \quad ...(3)$$

where F' gives the remaining degrees of freedom of the system. Equation (3) is known as *condensed phase rule* which is similar to the equation (2). Solid-liquid equilibria of number of types are known. But the following types are more important:

Type I. Two components are completely miscible m liquid phase and the solid phase consist of pure components. Examples are : (a) Lead-antimony, (b) Lead-silver, (c) Potassium iodide-water system.

Type II. The components enter into chemical combination giving rise to one or more compounds. Such types of two-component systems are of two kinds :

IInd (A). Two components form a compound with congruent melting point. Examples are :

(i) Zinc-magnesium system, (ii) $FeCl_3$–water system.

IInd (B). Two components form a compound with incongruent melting point. Examples are :

(i) Picric acid-benzene system, (ii) Gold-antimony system.

(ii) Sodium sulphate-water system.

Simple Eutectic System or Solid-liquid Equilibrium of Type I.

Consider a two-component system having two components A and B. These components are completely miscible in liquid state. *They do not form any compound.* The phase equilibrium diagram for this type is shown in Fig. 2.3.

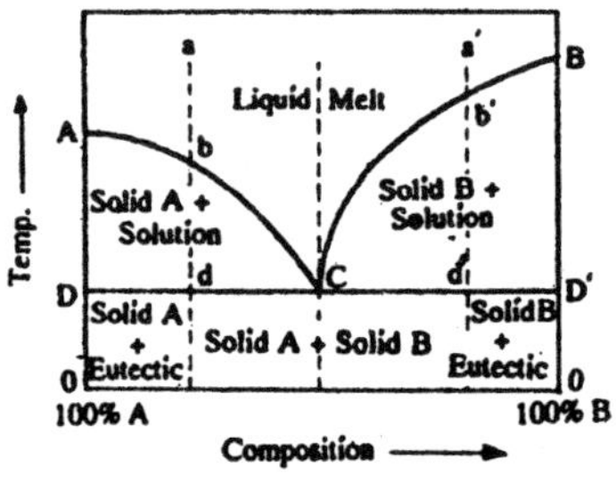

Fig. 2.3 : Solid-liquid system (Type I)

The phase diagram (3) consists of

(i) Curves. AC and CB

(ii) Points. A, B and C

(iii) *Areas* : ADC, D'CB, OO'D'D and they are above the curve ACB.

Description

(i) *Point A* : It represents the melting point of the pure component A.

(ii) *Point B :* It represents the melting point of the pure component B.

(iii) *Curve AC :* On gradual addition of small amounts of B to A, the freezing point of A is lowered along this curve AC. Therefore, this curve AC is known as the freezing point of the component A.

Along this curve AC, the solid A is in equilibrium with the solution of the component B in A (liquid) at different temperatures. Thus, the number of phases along this curve AC is two. As measurements are being made at constant atmospheric pressure, it follows from the condensed phase rule (3),

$$F = C - P + 1 = 2 - 2 + 1 = 1$$

Thus any point on the curve AC has one degree of freedom or is univariant. It means that only composition varies along the curve AC.

Curve BC : As increasing quantities of A are added to B, the freezing point of B falls along the curve BC. Along this curve, the solid B is in equilibrium with the solution of the component A in B at different temperatures. Again the number of phases along the curve BC like the curve AC is two. Again, applying the condensed phase rule equation, we get

$$F' = C - P + 1 = 2 - 2 + 1 = 1$$

Thus, any point on the curve BC is univariant, *i.e.,* only composition varies along BC.

Point C : At the point C, where the two curves AC and BC intersect, obviously both the sides A and B must be in equilibrium with the liquid phase. As three phases co-exist at this point, it follows from the modified phase rule (3).

$$F' = C - P + 1 = 2 - 3 + 1 = 0$$

Thus, the system at the point C is non-variant, *i.e.,* has no degree of freedom. This means that there is only one temperature at which the liquid phase can be in equilibrium with both solids A and B, and that the composition of the liquid phase must also be definite. From the figure, it is seen that the point C is the lowest temperature at which the liquid can exist. Therefore C has been called the eutectic point because it represents the lowest melting point of any mixture of solids A and B.

Area Above ABC : In this area the two components A and B exist as a homogeneous liquid solution because the temperature in it is above the melting point of any mixture. It means that there is only one phase in the area. Therefore, applying the condensed phase rule into this area, we obtain

$$F' = C - P + 1 = 2 - 1 + 1 s = 2$$

Thus, the system in the area above curve ACB is bivariant. It means that temperature and composition are required to define any point in the area above curve ACB.

Area OO'D'D : In this area only solid can exist because the liquid phase can not exist below the eutectic temperature.

Area ADC : Any point in the area ADC represents equilibrium between solid A and liquid. The composition of the phases in equilibrium can he obtained by drawing a horizontal line through the particular point. This line, where it cuts the curve AD, gives the nature of the solid and the AC to give the composition of the liquid.

Area CD'B : The above explanation is also true for this area.

Cooling of liquid mixtures : It is possible to predict the behaviour of any system on heating or cooling by using equilibrium diagram (3). This type of study is of special importance in the study of alloys in metallurgy.

Consider a liquid mixture of components represented by a point a. Allow it to cool at constant pressure. This fall in temperature continues without any change of composition till the point b on the curve AC is reached. At the point b, solid A separates out at a temperature t_1 which corresponds to the point b. Thus, the system at b involves two phases, *i.e.*, solid A and liquid solution.

Applying the condensed phase rule to this point b, it gives

$$F' = C - P + = 2 - 2 + 1 - 1$$

Thus, the system at point b is univariant. It means that there will be a fall in temperature only when the composition, of liquid phase changes. Therefore, if the solid A keeps on separating out along the curve bC and the solution becomes relatively richer in B. At the point y solid A is in equilibrium with its mixture of composition y at a temperature t_2 which corresponds to this point. Thus, in the area ACD solid A is in equilibrium with solution of different compositions depending upon different temperatures. When the point d is reached which corresponds to the eutectic point C, the second solid B also starts separating out. Now the system will have three phases. Again, applying the phase rulc to this system, we obtain

$$F' = C - P + 1 = 2 - 3 + 1 \text{ or } F' = 0$$

Thus, the system is univariant. It implies that on cooling further, the solid A and solid B separate out in the fixed ratio so that the composition of the liquid mixture remains constant as required at the point C. Here temperature does not change. This continues till the whole solution phase has been completely solidified. Thus, the system in the area OO'D'D consists of mixture of solid A and solid B. In that area, the system becomes univariant,

$$F' = C - P + 1 = 2 - 2 + 1$$

or $F' = 1$.

It means that on further cooling the temperature will only fall in the area BD'DA keeping the composition of two solids as constant. Similarly, one can explain if one considers a liquid of composition a' on the right side of eutectic point. Examples are : (i) Animony-lead system, (ii) Silicon-aluminium system. (Hi) Bismuth-cadmium system, (iv) KCl–$CaCl_2$ system, (v) Benzene-methyl chloride system, (vi) KI-H_2O system, (vii) Lead-silver system.

We will now discuss some examples of type 1 of two-component involving solid-liquid equilibria.

Lead-silver system : It is a two component system. The two metals lead and silver are completely miscible in the liquid state and do not form any compound. Therefore, the equilibrium diagram of this system is similar to that shown in the Fig. (2.4). The special features of lead-silver system are demonstrated in Fig. 2.4.

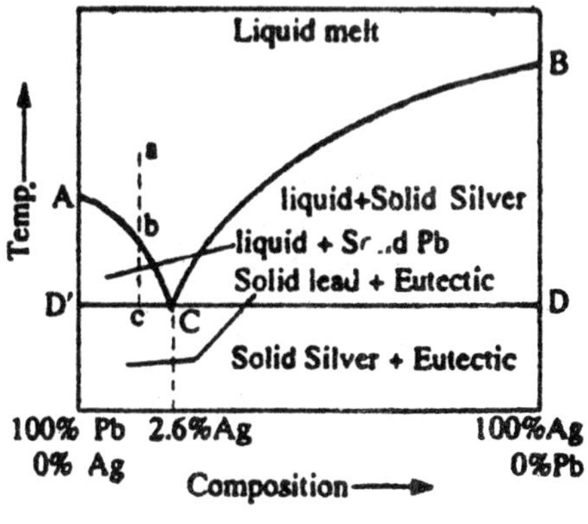

Fig. 2.4 : Lead-Silver System.

As the experiment is being carried out at constant pressure, the vapour phase is not considered and the condensed phase rule is applied.

$$F' - C - P + 1$$

Curve AC : A is the melting point of lead (327°). When increasing quantities of pure silver are added to lead metal, the freezing point of lead lowers along the curve AC. Therefore, AC is the freezing point curve of lead.

Along the curve AC, the solid lead is in equilibrium with the liquid melt. Thus, the number of phases along the curve is two. As all measurements are made at atmospheric pressure, on applying condensed phase rule, $F' = C - P + 1 = 2 - 2 + 1 = 1$

Therefore, the system is univariant. It means only composition varies along the curve AC.

Curve BC : Point B represents the melting point of pure silver (961°). When increasing quantities of pure lead are added, the freezing of silver lowers along the curve BC. Therefore, this curve may be called the *freezing point* curve of silver. Again, the solid silver is in equilibrium with the liquid melt along the curve BC. Thus, there are two phases. Applying the condensed phase rule equation, we obtain

$$F' = C - P + 1 = 2 - 2 + 1 = 1.$$

Again, the curve BC like AC is univariant and only composition varies along this.

Eutectic point C : The two curves AC and BC intersect at the eutectic point C. This point, being common to both the curves, represents the condition under which three phases lead, solid silver and liquid melt co-exist. The degree of freedom here is zero which also follows from reduced phase rule :

$$F' = C - P + 1 = 2 - 3 + 1 = 0$$

Thus, the point C is invariant, *i.e.,* has no degree of freedom. This point C (308°C) lies at a temperature which is lower than the melting points of silver as well as lead metal. The composition of an alloy of lead (97.6%) and silver (2.4%) corresponding to the point C is also fixed.

Therefore, the point C is called the *eutectic point i.e., lowest temperature at which the liquid melt can exist. The temperature corresponding to this point is known as eutectic temperature and composition to this point is known as eutectic composition.*

Area above curves AC and BC : In this area, silver and lead are present as a homogeneous liquid solution. Thus, there is only one phase in this area. Applying the condensed rule to any point in this area, we obtain :

$$F' = C - P + 1 = 2 - 1 + 1 = 2$$

Thus, the system in this area is bivariant. It means that both temperature and composition are required to define any point in this area.

Area ACD': In this area, solid lead is in equilibrium with the liquid melt.

Area BCD : In this area, solid silver is in equilibrium with the liquid melt,

Significance of lead-silver system : The phase diagram is utilised in the separation of silver from lead in Pattinson's Process for desilverisation of lead.

Suppose a represents the molten argentiferous lead containing a small amount of silver in it. Allow it to cool. Only temperature of the liquid melt falls with no change in concentration till the point b on the curve AC is reached. On further cooling lead begins to crystallise out and the solution becomes richer in silver. Further cooling will continue along the curve bC. Lead continues to separate out and is constantly removed by means of ladels. The liquid melt continues to be richer and richer in silver till the point C is reached where an alloy containing 2.6% of silver is obtained. This is then subjected to cupellation process.

The above process of increasing the concentration of silver in argentiferous lead is known as *Pattinson's process*.

General Discussion

(i) In Fig. 3.14 it is shown that the melting point (congruent) lies above the melting points of A and B. But it is not so. Certain cases are known in which the congruent melting point of the compound AB may be above, below or between those of two components A and B.

(ii) In Fig. 3.14 it is shown that the compound AB contains equimolecular amounts of A and B. This is not always so.

(iii) Draw a line XX_1. It means that the liquid phase has two compositions X and X, in equilibrium with the same solid AB at a certain temperature, say, t. In simple words it means that the compound AB has two solubilities at the same temperature* This paradox is explained by the diagram (2.4). AB is made up of two parts with DD' as the dividing line.

Thus, the left half of the diagram will be a two component system of solid A and compound AB and the right half of the diagram shows the two component system involving solid B and compound AB. Also, the curve DC will represent the freezing point curve of solid compound AB when A is added to it. Similarly, the curve DE will represent the freezing point curve of compound AB when B is added to it.

(iv) When two components form more than one compound, the equilibrium diagram will have curve analogous to CDE for each compound.

Examples are:

(i) Zinc-magnesium system,

(ii) Aluminium-magnesium system,

(ii) Mercury-thorium system,

(iv) Gold-tin system,

(v) Phenol-aniline system,

(vi) Ferric chloride-water system.

We will now discuss some examples.

Zinc-Magnesium System : It is an example of a two component system in which two metals form a compound having congruent melting point. The compound is an alloy having the formula $Mg(Zn)_2$ with a true melting point 590°C which lies between the melting points of two metals. The composition of $Mg(Zn)_2$ by weight is

$$Mg = 79\%$$

and Zn = 21%

The melting point of zinc is 420°C and that of magnesium is 650°C. The melting point of $Mg(Zn)_2$ is 590°C. The compound is fairly stable and it melts without change of composition.

As the experiment is being carried out at constant pressure, vapour phase therefore, is not considered and the condensed phase rule will be applied :

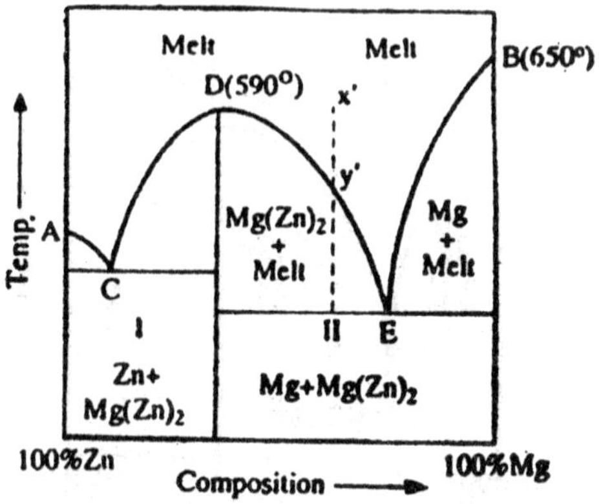

Fig. 2.5 : Zn—Mg system

$$F' = C - P + 1$$

The equilibrium diagram of such a system is represented in Fig. 2.5.

Point A : It represents the melting point of pure zinc (M.P. of zinc = 420°C],

Curve AC : By gradual addition of Mg to Zn, the melting point of zinc is lowered along the curve AC. Along this curve the solid zinc is in equilibrium with melt (solution). Thus, the number of phases is two. Applying the condensed phase rule equation to this curve, we get

$$F' = C - P + 1 = 2 - 2 + 1 = 1$$

Thus, the system is univariant, *i.e.,* only the composition varies along the curve.

Point B : This represents the melting point of pure magnesium [M.P. = 65°C].

Curve BE : When zinc is added to Mg the melting point of Mg is lowered along the curve BE. Along the curve BE, the solid Mg is in equilibrium with melt at different temperatures and hence is regarded as the freezing or melting point curve of magnesium. Again, only composition varies along this curve which is evident from phase rule,

$$F' = C - P + 1 = 2 - 2 + 1 = 1 \qquad [\because P=2]$$

Curve DC : When further quantities of zinc are added, the melting point of the compound Mg $(Zn)_2$ lowers along this curve. Thus, the curve represents the lowering of melting point of compound Mg $(Zn)_2$ when zinc is added.

Curve DE : This curve represents the lowering of melting point of compound Mg $(Zn)_2$ when magnesium metal is added to it.

Point D : At this point, the liquid and solid have the same composition, *i.e.,* $Mg(Zn)_2$ and melt at this point without decomposition. Thus, the temperature corresponding to the point D is regarded as-the congruent melting point of $Mg(Zn)_2$.

Point C : The two curves AC and CD intersect at the point C. Thus, there will be three phases at this point. These three phases are solid zinc, compound $Mg(Zn)_2$ and melt. This point is invariant which is evident from the phase rule,

$$F' = C - P + 1 = 2 - 3 + 1 = 0$$

The point C is known as eutectic point.

Point E : This is the point where two curves BE and DE intersect at the point E. Thus, there are three phases which are in equilibrium at this point. These three phases are solid Mg, compound $[Mg(Zn)_2]$ and melt. Again, it is invariant and is known as eutectic point.

Cooling : Consider a melt of composition represented by a point x. Allow it to cool along the line xy. When it reaches y, the solid $Mg(Zn)_2$ separates out from the melt. At this point there are two phases and hence the system becomes univariant.

$$F' = C - P + 1 = 2 - 2 + 1 = 1$$

It means that the change in composition will also take place if the temperature is further lowered beyond the point y. This cooling will, therefore, take place along yC. At the point C, solid zinc will separate out. Thus, there are three phases at this point and hence it is nonvariant.

$$F' = C - P + 1 = 2 - 3 + 1 = 0$$

It means that there will not be any change in temperature as well as composition at this point if cooling is continued and as long as all the three phases are present. Similarly, consider a liquid melt of composition represented by a point x'. Allow it to cool. The temperature will, fall along x'y'. At point y'. solid $Mg(Zn)_2$ separates. Further cooling will therefore, take place along y 'E'. At the point E, magnesium also separates out.

THE SULPHUR SYSTEM

It is a one component system consisting of four phases :

Rhombic sulphur S_R...............Solid phase

Monoclinic sulphur S_M. ....Solid phase

Liquid sulphur S_L,.................. Liquid phase

Vapour sulphur S_V..................Vapour phase.

The phase diagram of sulphur system is represented.

The diagram consists of :

(i) *Curves* AB, BC, CD, BE, CE and EG and for metastable curves.

(ii) *Areas* : The above curves divide the diagram into 4 areas ABEG, ABCD, GECD and BEC.

(iii) *Stable triple points* : Three stable triple points are B. C and E.

(iv) *Metastable point* : F is metastable point.

Description

Curve AB : This curve is the *sublimation* or *vapour pressure* curve of rhombic sulphur (S_R). This gives the variation of vapour pressure of S_R at different temperatures. Thus, along this curve, two phases rhombic sulphur and vapour are in equilibrium with each other. It follows from the phase rule.

$$F' = C - P + 2 = 1 - 2 + 2 = 1.$$

Thus, the system is univariant. It means that there can be one vapour pressure value at one temperature.

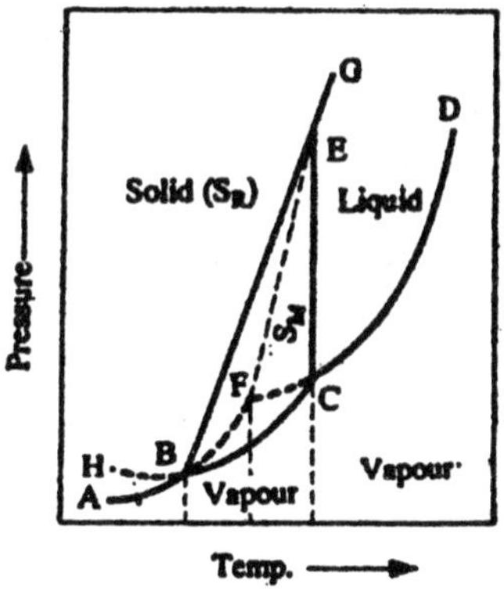

Fig. 2.6 : Sulphur system.

Point B : At the point B, rhombic sulphur changes into monoclinic sulphur. This is known as transition temperature. Thus, there are three phases (two solids and one vapour) in equilibrium at this point (B) and it is non-variant.

$$F = C - P + 2 = 1 - 3 + 2 = 0$$

Thus, B is a triple point.

Curve BC : It is the *vapour pressure* or *sublimation* curve for monoclinic sulphur. It gives the variation of vapour pressure of S_M (monoclinic sulphur) at different temperatures. At any point on this curve BC, the two phases SM and Sy are in equilibrium. The system is univariant. *i.e.*, it has one degree of freedom.

$$F = C - P + 2 = 1 - 2 + 2 = 1$$

Point C : It represents the *melting point* of monoclinic sulphur (120°). This is another triple point. At this point three phases (S_M. S_L, and S_V) are in equilibrium. Thus, this is a non-variant point.

Curve CD : It is the vapour pressure curve of liquid sulphur. This curve starts at point C which is melting point of S_M and terminate at point D, the critical temperature. Beyond this temperature there is one phase S_V.

Along this curve SL and, Sy are in equilibrium. Thus the system is univariant since at any point on the curve for the fixed value of pressure, temperature has a fixed value. This is also evident from the phase rule,

$$F = C - P + 2 = 1 - 2 + 2 \text{ or } F - 1$$

Curve BE : This is known as the *transition curve* of S_R to S_M. It gives the effect of pressure on the transition point. Both phases (S_R and S_M) are solids. The system is monovariant. This curve terminates at E beyond which SM disappears. As the transition point is raised With increase of pressure, the curve BE, therefore, slopes away from the pressure axis.

Curve CE : It is the melting or fusion curve for monoclinic sulphur. It shows the effect of pressure on melting point of S_M which is raised by increase of pressure, the curve CE, therefore slopes slightly away from the pressure axis.

At any point of the curve, the two phases S_M and S_L are in equilibrium with each other. This also has one degree of freedom.

Point E : The two curves BE and CE meet at the point E. Thus, it is another triple point which is nonvariant. At this point, three phases (S_R, S_M and S_L) are in equilibrium.

Curve EG : It is the fashion, curve for rhombic sulphur. Along this curve S_R and S_L are in equilibrium. Therefore, the number of phases is one and hence the system is monovariant.

Metastable Equilibrium

Curve BF : If the temperature of rhombic sulphur at 95.5°C is allowed to rise quickly, the transition of S_R or S_M does not take place at B But this curve is extended to F, the melting point of S_R. The curve BF so obtained is the metastable curve of rhombic sulphur. If system is disturbed slightly, the curve BE merges into BC.

Curve CF : If liquid sulphur is allowed to cool very carefully the solid phase will not separate out at C. Thus, the curve CD can be extended to F by supercooling the liquid carefully. Thus, the curve CF represents metastable equilibrium between S_L and S_V.

Curve EF : It is the metastable fusion curve of rhombic sulphur. Along this curve, rhombic sulphur is in metastable equilibrium with liquid sulphur. This curve gives the effect of pressure on the melting point of rhombic sulphur in metastable state.

Point F : It is the metastable triple point where S_R. S_L and S_V co-exist in the equilibrium ; like B, C and E, it is non-variant.

Areas. There are four areas :

(i) Area ABEG : It contains only rhombic sulphur.

(ii) Area ABCD : It contains only vapour sulphur.

(iii) Area GECD : It contains on ly liquid sulphur.

(iv) Area BEC : It contains only monoclinic sulphur.

Thus, all the above areas contain one phase only. It means that there are two degrees of freedom which is also evident from the phase rule equation.

$$F = C - P + 2 = 1 - 1 + 2 \text{ or } F = 2$$

It means that both pressure and temperature are required to define any point in all the above areas.

SOLUBILITY PRODUCT

Consider the *saturated* solution of a sparingly soluble binary electrolyte AB. There exist two equilibria simultaneously between undissolved solid, dissolved (but unionised) molecules and free ions, *i.e.*,

$$\underset{\text{(Undissolved solid)}}{AB} \underset{}{\overset{K_1}{\rightleftharpoons}} \underset{\text{(Dissolved, but unionsed)}}{AB} \underset{}{\overset{K_2}{\rightleftharpoons}} \underset{\text{(Ions in solution)}}{A^+ + B^-}$$

Applying the law of mass action, it follows that

$$\frac{[A^+][B^-]}{[(AB)_d]} = K_1; \qquad \frac{[(AB)_d]}{[(AB)_d]} = K_2$$

∴ On multiplying, we get

$$\frac{[A^+][B^-][AB)_d]}{[(AB)_d][AB)_s]} = \frac{[A^+][B^-]}{[(AB)_s]} = K_1 \times K_3$$

But the concentration of solid AB is constant, so that

$$[A^+][B^-] = \text{a constant} = K_s$$

where the constant K_s is known as the solubility product of AB.

In general, for an electrolyte, A_xB_y, which ionises as

$$A_xB_y \rightleftharpoons xA^{y+} + yB^{x-}$$

the solubility product is given by

$$K_s = [A^{y+}]^x [B^{x-}]^y$$

∴ where $[A^{y+}]$ and $[B^{x-}]$ are the ionic concentration in a *saturated solution*. Thus, solubility product of an electrolyte may be defined as the *maximum product of the concentrations of its constituent ions (expressed in g-ion per litre) in its solution, when each ionic concentration term being raised to the number of times the ion occurs in the equation representing the solution of 1 molecule of the electrolyte.* For example for

$$Ag_2CrO_4 \rightleftharpoons 2Ag^+ + CrO_4^{2-}$$

the solubility product, $K_s = [Ag^+]^2[CrO_4^{2-}]$.

Relation Between Solubility and Solubility Product

Consider the general form of electrolyte A_xB_y which dissociates as

$$\underset{S}{A_xB_y} \rightleftharpoons \underset{xS}{x[X^{y+}]} + \underset{yS}{y[X^{x-}]}$$

Let S gram mole/litre be the solubility of the electrolyte, then

$$[A^{y+}] = x\,.\,S \quad \text{and} \quad [B^{x-}] = y\,.\,S$$

$$\therefore \qquad K_s = [A^{y+}]^x[B^{x-}]^y$$

$$= [x . S]^x[y . S]^y = x^x . y^y . S^{x+y}$$

For example in case of $BaCl_2 \rightleftharpoons Ba^{2+} + 2Cl^-$; x = 1 and y = 2

$$Ks \rightleftharpoons 1^1 . 2^2 S^{1+2} = 4S^3.$$

Analytical Applications of Solubility Product

From the relation:

$$K_s = [A^+]^x[B^-]^y$$

it is clear that:

(a) a solution in which $[A^+]^x[B^-]^y$ is *less than* K_s, the solution is unsaturated and more A_xB_y can be dissolved in it,

(b) a solution in which $[A^+]^x[B^-]^y$ is *equal* to K_s, *is just saturated*, and

(c) if anything is done to the solution which tends to make $[K^+]^x [B^-]^y$ *greater than* K_s, then solid A_xB_y will be precipitated.

From the above discussion it is evident that if the product of concentration of ions in a saturated solution of an electrolyte becomes greater than the solubility product of the electrolyte, it will precipitate out. In qualitative analysis, separation and identification of metal ions such as Cu^{2+}, Zn^{2+}, etc., is possible only due to difference in solubility products of their sulphides. The above concepts of solubility product have become extremely useful in getting a clear picture of different processes in analytical chemistry. A few cases are illustrated below:

1. Qualitative Analysis of Cationic Mixtures

The identification of individual species of cations present in an unknown mixture of electrolytes involves a preliminary segregation of the cations into selected classifiable groups.

A simple system devised for this purpose is based on the selective precipitation of sparingly soluble compounds of the cations from aqueous solution of the electrolyte mixture.

In its most simplified form, for the more common cations only, these compounds fall into five groups. Within each group the anion is the same for the different cations and the solubility products are small and do not differ too greatly from one another:

Group I: The sparingly soluble chlorides derived from Pb^{2+}, Hg_4^{2+} and Ag^+ ions are precipitated by the addition of dilute HCl.

Chloride	*K_s at 25°C*

$PbCl_2$	1.6×10^{-5}
Hg_3Cl_2	1.3×10^{-15}
AgCl	1.8×10^{-16}

Group II: The sparingly soluble sulphides derived from Hg^{2+}, Pb^{2+}, Cu^{2+}, Cd^{2+} and Bi^{3+} ions are precipitated by the passage of H_2S through an acidic solution whose concentration of hydrogen ions is small.

Sulphide	*K_s at 25°C*
HgS	3×10^{-32}
PbS	8×10^{-28}
CuS	8×10^{-36}
CdS	7×10^{-27}
Bi_2S_3	1×10^{-96}

Group III: The sparingly soluble hydroxides derived from Fe^{2+}, Cr^{3+} and Al^{3+} ions are precipitated by the addition of ammonium hydroxide solution in the presence of excess ammonium chloride

Hydroxide	*K_s at 25°C*
$Fe(OH)_3$	6×10^{-28}
$Cr(OH)_3$	7×10^{-31}
$Al(OH)_3$	1.4×10^{-24}

Group IV: A further group of sparingly soluble sulphides, those derived from CO^{2+}, Ni^{2+}, Mn^{2+} and Zn^{2+} ions are precipitated by the passage of H_2S through a basic solution.

Sulphide	*K_s at 25°C*
CoS	8×10^{-23}
NiS	2×10^{-21}
MnS	1×10^{-11}
ZuS	8×10^{-25}

Group V: The sparingly soluble carbonates derived from Ca^{2+}, Sr^{2+} and Ba^{2+} ions are precipitated by the addition of ammonium carbonate solution in the presence of NH_4OH and NH_4Cl.

Carbonate	*Ks at 26°C*
$CaCO_3$	4.7×10^{-9}

$SrCO_3$	7×10^{-10}
$BaCO_3$	1.6×10^{-9}

The most common cations remaining in solution after the precipitation fo group V are Na^+, K^+ (all the simple salts of the alkali metals are water-soluble) and Mg^{2+} ions. The order of group precipitation is such that the precipitating reagents for later groups also precipitate in most cases cations in preceding groups, which must therefore the completely removed in the above prescribed order if they are not to interfere. For example,

(i) K_S of silver sulphide is 7×10^{-50} at 25°C. It means that silver sulphide will be precipitated in group II unless it has been completely removed as AgCl in group I.

(ii) Any of the cations normally precipitated as their sulphides in group II, will reappear in the group IV precipitate unless completely removed in group II.

(ii) K_S for $ZnCO_3$ at 25°C is 2.1×10^{-11}. Hence $ZnCO_3$ will be precipitated in group V unless it has been completely removed as ZnS in group IV.

2. The Separation of the Sulphides into Groups II and IV

The values of K_S for group II sulphide are smaller than those for group IV sulphides. Hence, it is possible to precipitate the sulphides of those cations which fall in group II selectively in the presence of those cations which fall in group IV if the $[S^{2-}]$ from the reagent H_2S is regulated such that it exceeds that required to reach the solubility products of the group II sulphides but is less than that required to reach the solubility produces of the group IV sulphides. This is done as follows:

H_2S is a weak diprotic acid for which:

(a) $$H_2S \rightleftharpoons H^+ + HS^-$$

$$\frac{[H^+][HS^-]}{[H_2S]} = K_{S_1} = 9.1 \times 10^{-3} \text{ at } 18°C$$

(b) $$HS^- \rightleftharpoons H^+ + S^{2-}$$

$$\frac{[H^+][S^{2-}]}{[HS^-]} = K_{S_2} = 1.2 \times 10^{-15} \text{ at } 18°C$$

Hence, on multiplying:

$$\frac{[H^+]^3[S^{2-}]}{[H_2S]} = 10^{-22}$$

$$\therefore \qquad [S^{2-}] = \frac{[10^{-22}][H_2S]}{[H^+]^2}$$

If H_2S at 1 atmosphere is bubbled through water, for the saturated solution:

$$[H_2S] \sim 0.1 \text{ mole/litre}$$

$$\therefore \qquad [S^{2}] = \frac{10^{-22} \times 10^{-1}}{[H^+]^2} = \frac{10^{-22}}{[H^+]^2}$$

i.e., the $[S^{2-}]$ is dependent on the $[H^+]$ and may be varied by varying the acidity of the solution.

If $\quad pH = 0(1\ MHCl)$

$\therefore \quad [H^+] = 1$ mole/litre

or $\quad [S^{2-}] = 10^{-22}$ mole/litre

If $\quad pH = 12(0.01\ M\ NaOH)$

$\therefore \quad [H^+] = 10^{-12}$ mole/litre

or $\quad [S^{2-}] = 10$ mole/litre

Hence if the acidity is too low, the corresponding $[S^{2-}]$ is large enough to exceed the requirements of the solubility products of both the group II and the group IV sulphides which are precipitated together in group II.

Conversely, if the acidity is too high, the $[S^{2-}]$ ion is suppressed to a degree when it is lower than that required to reach the solubility products of even the group II sulphides. These, therefore, remain in solution and are precipitated in group IV where the $[S^{2-}]$ in the basic solution is too high.

In practice, if the $[H^+]$ of the solution, after the precipitation of the group I chlorides is adjusted to 0.3 M prior to passing H_2S, the group II sulphides are precipitated selectively.

(a) Selective precipitation of the group III hydroxide: Three hydroxides only are precipitated in group III, those of Fe^{2+}, Cr^{3+} and Al^{3+} whose K_s values lie between 10^{-30} to 10^{-33}. However, the K_s values for the hydroxides of those cations which are normally precipitated in group

IV as their sulphide are also relatively small, though of a higher order than the above there:

Hydroxide	K_s *at 25°C*
$Co(OH)_2$	2×10^{-16}
$Ni(OH)_2$	2×10^{-15}
$Mn(OH)_2$	1.6×10^{-13}
$Zn(OH)_2$	7×10^{-18}

Now, for a solution fo NH_3 in water the following equilibria exist:

$$NH_3 + H_2O \rightleftharpoons NH_4OH \rightleftharpoons NH_4^+ + OH^-$$

for which $\dfrac{[NH_4^+][OH^-]}{[NH_4OH]} = K_b = 1.8 \times 10^{-5}$ at 25°C

$\therefore \quad [NH_4^+][OH^-] = 1.8 \times 10^{-5}\,[NH_4OH]$

But $\quad [NH_4^+] = [OH^-]$

$\therefore \quad [OH^-]^2 = 1.8 \times 10^{-5}\,[NH_4OH]$

For 1 M NH_4OH :

$$[OH^-]^2 = 1.8 \times 10^{-5}$$

or $\quad [OH^-] = 4.24 \times 10^{-3}$ mole/litre

For 0.1 M NH_4OH:

$$[OH^-]^2 = 1.8 \times 10^{-5} \times 10^{-1} = 1.8 \times 10^{-6}$$

or $\quad [OH^-] = 1.34 \times 10^{-2}$ mole/litre

Hence, if a solution of NH_4OH alone is used to precipitate the hydroxides in group III, the $[OH^-]$ is high enough to exceed the requirements of the solubility products for the hydroxides of the cations which fall in group IV. These are therefore precipitated together with the normal group III hydroxides.

It is therefore necessary to suppress the $[OH^-]$ concentration to such an extent where it exceeds the requirements of the solubility products for the group III hydroxides but is lower than is necessary to reach the solubility products for the hydroxides of the group IV cations. This is done by the addition of an excess of NH_4Cl, when a buffer solution is produced consisting of the weak base, NH_4OH, and a salt of the weak base and a strong acid, NH_4Cl. NH_4Cl, being a salt, is a strong electrolyte:

$$NH_4Cl \rightarrow NH_4^+ + Cl^-$$

The excess NH_4^+ ions depress the ionisation of NH_4OH in order that K_b for NH_4OH may be maintained. Hence the $[OH^-]$ is suppressed to a point where only the group III hydroxides are precipitated.

(b) Quantitative Analysis: The solubility products of the silver halides are small:

At 25°C, K_s of AgCl $= 1.8 \times 10^{-10}$

K_s of AgBr $= 5.2 \times 10^{-13}$

K_s of AgI $= 8.3 \times 10^{-17}$

Hence, if a solution fo the soluble $AgNO_3$ is added to a solution containing a soluble halide, the halide ions are quantitatively removed from solution as the sparingly soluble silver salt. The concentration of halide ions remaining in solution when precipitation is just complete is very low and may be neglected.

For example, if $AgNO_3$ solution is added to NaCl solution, AgCl is precipitated. When precipitation is just complete, for the ions remaining in solution:

$$[Ag^+][Cl^-] = K_s = 1.8 \times 10^{-10}$$

But $$[Ag^+] = [Cl^-]$$

$\therefore$ $$[Cl^-] = (1.8 \times 10^{-10})^{1/2} = (1.34 \times 10^{-5}) \text{ mole/litre}$$

Thus precipitation of sparingly soluble silver halides is made the basis of quantitative determinations of the concentrations of solution of soluble halides. A solution of the halides of unknown concentration is titrated with a standard solution of $AgNO_3$. The end point of the reaction occurs when precipitation fo the sparingly soluble silver halide is just complete.

3. Using Excess of Reagent

The use of excess of reagent is essential to ensure complete precipitation in gravimetric analysis. The insoluble substances like AgCl, $BaSO_4$, etc., have very slight solubility but by adding excess of precipitant their solubility product is further exceeded. Thus, the ionic product $[Ba^{2+}][SO_4^{2-}]$ becomes still higher than the solubility product of $BaSO_4$, by using large excess of H_2SO_4.

$$BaSO_4 \rightleftharpoons Ba^{2+} + SO_4^{2-}$$

$$H_2SO_4 \rightleftharpoons 2H^+ + SO_4^{3-}$$

4. Use of K_2CrO_4 as Indicator in $AgNO_3$ Chloride Titrations

In silver nitrate-chloride titration, the $AgNO_3$ is added from the burette into the chloride solution to which some K_2CrO_4 solution is added. At first the added $AgNO_3$ precipitates AgCl, but not Ag_2CrO_4 because AgCl has a lower solubility (0.00145 g per litre) than Ag_2CrO_4 (0.0285 g per litre). When sufficient $AgNO_3$ has been added to precipitate all the AgCl, subsequent addition causes the formation of brick-red Ag_2CrO_4 precipitate, which serves to note the end-point.

IONIC PRODUCT OF WATER

Pure water is essentially a covalent compound. Nevertheless, it ionises very slightly and the following equilibrium is established:

$$H_2O \rightleftharpoons H^+ + OH^- \qquad ...(1)$$

where H^+ is a hydrogen ion and OH^- is a hydroxyl ion. The removal fo the one orbital electron from a hydrogen atom yields the positive hydrogen ion or hydrogen cation, which is in fact a bare proton. The bare proton has no separate existence in solution and owes its stability to solvation by a water molecule to give the hydronium ion, H_3O^+. The ionisation equilibrium of water is thus represented more accurately as:

$$2H_2O \rightleftharpoons H_3O^+ + OH^-$$

However, as many qualitative and quantitative considerations remain essentially the same for the unsolvated and the solvated proton, it is usual for convenience to refer to the unsolvated proton, H^+.

On applying the law of mass action to equation (1), we get

$$K = \frac{[H^+][OH^-]}{H_2O} \qquad ...(2)$$

where, K is called the ionisation constant. As water is ionised to very slight extent (about 1 in 10^7 water molecules), the concentration of water molecules is very large as compared with H^+ and OH^- ions so that it can be regarded as practically constant.

Therefore, equation (2) may be written as

$$[H^+][OH^-] = K[H_2O] = k_w \qquad ...(3)$$

where constant k_w is constant called the *ionic product of water.*

Conductivity measurements show that in pure water,

$$[H^+] = 10^{-7} \text{ mole/litre at } 25°C$$

Furthermore in pure water,

$$[H^+] = [OH^-]$$

Therefore, equation (3) may become as

$$[10^{-7}][10^{-7}] = k_w$$

or $$k_w = 10\text{– }14 \text{ at } 25°C \quad ...(4)$$

Significance of Equation (3) and (4)

For all aqueous solutions, the product of $[H^+][OH^-]$ may be taken to be equal to 10^{-14}. If extra H^+ ions are introduced, the ionisation of H_2O is suppressed to the point at which $[H^+][OH^-] = 10^{-14}$. Similarly, if extra OH^- ions are introduced, the ionisation of water is again suppressed to maintain the constancy of $[H^+][OH^-]$. From equation (3), it follows that the ionic product of water may be defined as *the product of concentration of H^+ and OH^- ions expressed in gram ions per litre.*

TRANSPORT NUMBER

The fraction of the total current carried by any one of the ionic species is known as transport number, transference number or Hittorf number of the species and may be denoted by sets of symbols like t_+ and t_-, t_a and t_c or n_c are n_a.

From the above definition, it follows that the transport number of an anion is

$$t_- = \frac{\text{Current carried by an anion}}{\text{Total current passes through the solution}} \quad ...(1)$$

and the transport number of cation is

$$t_+ = \frac{\text{Current carried by the cation}}{\text{Total current passed through the solution}} \quad ...(2)$$

But we know that the quantity of electricity carried by each ion is directly proportional to the speed of the concerned ions.

So, amount of electricity carried by anion $\propto$ speed of anion (u_a) amount of electricity carried by cation $\propto$ speed of cation (u_c).

$\because$ Total amount of electricity carried $\propto$ speed of anion + speed of cation

$\propto u_a + u_c$

Thus, the transport number of anion [From eq. (1), t_- is]

$$t_- = \frac{u_a}{u_a + u_c} \qquad ...(3)$$

and the transport number of cation [From eq. (2)], t_+, is

$$t_+ = \frac{u_c}{u_a + u_c} \qquad ...(4)$$

It follows from equation (3) and (4) that the transport number of an ion may be defined as the ratio of speed of the ion to the sum of the speeds of both cation and anion.

By adding Eqs. (3) and (4), we get

$$t_- + t_+ = \frac{u_a}{u_a + u_c} + \frac{u_c}{u_a + u_c} = \frac{u_a + u_c}{u_a + u_c} = 1 \qquad ...(5)$$

$$t_- + t_+ = 1$$

From eq. (5), it follows that

(i) The sum of the transport numbers of all the species taking part in the transport of electricity is equal to unity, and

(ii) If the transport number of one of the ionic species is known, then that of the other can be easily calculated by using eqn. (5).

Experimental Determination of Transport Number

There are several methods for determining the transport numbers of ions, viz.

(i) Hittorf's Method

Principle : This method is based on the principle that "*During electrolysis, the fall of concentration around the electrode is proportional to the speed of ions moving away from that electrode.*"

Thus, the fall of concentration around the cathode ∝ speed of the anion

(u_a), ...(6)

and the fall of concentration around the anode ∝ speed of the cation

(u_c) ...(7)

Dividing eq. (6) by (7), we get

$$\frac{\text{Fall of concentration around the cathode}}{\text{Fall of concentration around the anode}} = \frac{u_a}{u_c}$$

Adding 1 to both sides, we get

$$1 + \frac{\text{Fall of concentration around the cathode}}{\text{Fall of concentration around the anode}} = 1 + \frac{u_a}{u_c}$$

or
$$\frac{\text{Fall of concentration around the anode + Fall of concentration around the cathode}}{\text{Fall of concentration around the anode}} = \frac{u_c + u_a}{u_c}$$

or
$$\frac{\text{Total fall of concentration}}{\text{Fall of concentration around the anode}} = \frac{u_c + u_a}{u_c}$$

Taking reciprocals of the abode equation,

$$\frac{\text{Fall of concentration around the anode}}{\text{Total fall of concentration}} = \frac{u_c}{u_c + u_a} \quad \text{...(8)}$$

From equation (4), we know $t_+ = \frac{u_c}{u_a + u_c}$

$$\frac{\text{Fall of concentration around the anode}}{\text{Total fall of concentration}} = t_+ \quad \text{...(9)}$$

Similarly,

$$\frac{\text{Fall of concentration around the cathode}}{\text{Total fall of concentration}} = t_- \quad \text{...(10)}$$

Working

(i) A dilute solution fo known concentration (N/10) of the salt containing the ions, of which the transport numbers are to be determined, is filled in the apparatus.

(ii) The apparatus is then connected to a battery, variable resistance and voltameter in series as shown in Fig. (2.15). A low current of the order of 1 – 20 milliamperes is passed for 2 to 3 hours.

If high current density is used, diffusion sets in and the solution sin the various compartments are mixed and thus leading to wrong results.

(iii) After the electrolysis the stop cocks connecting the U-tube with the side tubes are closed and the solutions are withdrawn from the various compartments separately. The concentration is determined by titration against KCNS solution. These should be no change in the concentration of the solution from the V-tube, *i.e.*, central compartment. Knowing the original concentration of the solution taken, the change in concentration is determined separately.

(iv) The weight of Ag(or Cu) deposited in the voltameter is also determined separately. This represents the total quantity of electricity passed during electrolysis.

Calculations: Calculation involves the following steps:

Case I : *When the electrodes are unattackable.* This condition can be obtained by using platinum electrodes. Suppose that,

(i) The weight of $AgNO_3$ in the a g of anodic solution before electrolysis = X gm.

(ii) The weight of $AgNO_3$ in the b g of anodic solution after electrolysis = Y gm.

(iii) The weight of $AgNO_3$ in the a g of anodic solution after electrolysis

$$= \frac{Y}{b} \times a = Z$$

$\therefore$ Full in concentration around anode = (X – Z) gm.

Let the weight of Ag deposited in the voltameter = 2 gm

$$\frac{\text{Total weight of AgNO}_3\text{ electrolysed}}{\text{Weight of Ag deposited in voltameter}}$$

$$= \frac{\text{Eq. wt. of AgNO}_3}{\text{Eq. wt. of Ag}}$$

$$\therefore \quad \frac{\text{Total weight of AgNO}_3\text{ electrode}}{w} = \frac{170}{108}$$

or $$\text{Total weight of AgNO}_3\text{ electrolysed} = \frac{w \times 170}{108} = W \text{ (say)}$$

Now the transport No. of Ag^+ ion is given by

$$t_{Ag^+} = \frac{\text{Loss in wt. of AgNO}_3\text{ around anode}}{\text{Total weight of AgNO}_3\text{ electrolysed}}$$

$$= \frac{X - Z}{W} \qquad ...(11)$$

and the transport No. of NO_3^- ions is given by

$$t_{NO_3^-} = 1 - t_{Ag^+} = 1 - \frac{X - Z}{W} \qquad ...(12)$$

Case II : *When the electrodes are attackable.* This condition can be obtained by using Ag anode in $AgNO_3$ solution.

In this case $Z \geq X$, *i.e.*, the concentration of $AgNO_3$ in the anodic compartment after electrolysis is more than that before electrolysis. This is due to the fact that NO_3^- ions after losing charge at the anode attack the silver anode to form $AgNO_3$. Thus, the concentration of $AgNO_3$ in the anodic compartment increases.

The fall in concentration of $AgNO_3$ in the anodic compartment due to the migration of the Ag^+ ions may be calculated as follows:

The amount of $AgNO_3$ formed by the discharge of NO_3^- ions at the anode = No. of the NO_3^- ions discharged.

= total $AgNO_3$ electrolysed W (say)

$\therefore$ The actual increase in conc. of anodic compartment = W.

But observed increase in conc. of anodic compartment = Z – X.

$\therefore$ Fall in concentration due to the migration of Ag^+ ions

= Actual increase – observed increase

= W – (Z – X)

Hence $$t_{Ag^+} = \frac{W - (Z - X)}{W} \quad ...(13)$$

and $$t_{NO_3^-} = 1 - t_{Ag^+} = 1 - \frac{W - (Z - X)}{W} = \frac{Z - X}{W} \quad ...(14)$$

(2) Moving Boundary Method

Principle: It is based on measuring the rate of migration of one or both of the ionic species of the electrolyte, away from the similarly charged electrodes. This method was first of all suggested by Lodge and Whetham. It was modified by Masson (1899), Steel (1904) and Denison (1906). Finally it was the work of Macinnes (1923) whose work made the procedure more easy and simple.

In most of the methods, LiCl is not used as an indicator but instead of this cadmium is used as an anode which on electrolysis yields cadmium chloride which in turn itself acts as an indicator electrolyte. The electrolytic cell consists of the shape as shown, the middle part of the left hand portion is vertical tube of uniform area of cross section.

The experimental solution NaCl is floated on top of a denser solution of $CdCl_2$. At the bottom of the tube is fixed an anode of cadmium metal. Over the solution of NaCl a less dense solution of sodium acetate is floated. The cathode used is of platinum. This is placed in a side tube so that the evolution of hydrogen gas at the cathode does not disturb the main liquid column.

Working

(i) A constant current is passed for 5 to 6 hours. The Na^+ ions move upwards towards the cathode closely followed by the Cd^{2+} ions. The sharp boundary moves gradually upwards. The movement of the boundary between NaCl and $CdCl_2$ can be easily followed on account of a difference in the refractive indices of the two solutions.

(ii) The rate of migration of Cl^- ions can also be observed as the acetate anions are slow moving than the chloride ions.

(iii) At the end of experiment, the distance through which the boundary has moved is noted. The time for which the current has passed is noted.

Calculations: Suppose a current of I amperes is passed for t seconds, the quality of electricity carried by Na^+ ions is given by

$$= t_+ \, It \text{ coulombs}$$

where t_+ is the transport number of the Na^+ cation

Suppose the concentration of NaCl solution = c gm eq. eq/c.c.

Suppose the area of cross section of the moving boundary = a cm^2.

Therefore, the amount of Na^+ ion that has migrated upwards

$$= \frac{t_+ It}{F} \text{ where F is Faraday.} \quad ...(15)$$

Distance by which NaCl—$CdCl_2$ boundary moving during t seconds = x cm.

Volume of the solution cleared by the migration of Na^+ ions = ax

Number of gm equivalent of Na^+ ions transported = axc ...(16)

From Eqs. (15) and (16), we have $axc = \frac{t_+ It}{F}$

or

$$t_+ = \frac{F\,axc}{It} \quad ...(17)$$

Similarly,

$$t_- = \frac{F \cdot ayc}{It} \quad ...(18)$$

where y is the distance by which NaCl—CH_3COONa boundary moves during t send and t_- is the transport number of chloride anion.

Also $$\frac{t_+}{t_-} = \frac{x}{y}$$

But $$t_- = 1 - t_+$$

$\therefore$ $$\frac{t_+}{1-t_+} = \frac{x}{y} \quad \text{or} \quad t_+ = \frac{x}{x+y} \qquad ...(19)$$

Similarly $$t_- = \frac{x}{x+y} \qquad ...(20)$$

Thus, t_+ and t_- can be evaluated.

(3) From Ionic Mobility

The ionic mobility or conductance of an ion is directly proportional to the speed of the ion. Thus,

$$\lambda_+ \propto u_+$$

or $$\lambda_+ = Ku_+ \qquad ...(21)$$

Similarly, $$\lambda_- \propto u_-$$

or $$\lambda_- = Ku_- \qquad ...(22)$$

In Eqs. (21) and (22), u_+ and u_- are the speeds while λ_+ and λ_- are the ionic conductivities of the cation and anion respectively.

Now the transport number of cation is given by

$$t_+ = \frac{u_+}{u_+ + u_-}$$

$$= \frac{\frac{1}{K}\lambda_+}{\frac{1}{K}(\lambda_+ + \lambda_-)} \quad \text{[From Eqs. (21) and (22)]}$$

$$= \frac{\lambda}{\lambda_+ + \lambda_-} \qquad ...(23)$$

But $$\lambda_+ + \lambda_- = \lambda_\infty \text{ (Kohlrausch's Law)}$$

$\therefore$ $$t_+ = \frac{\lambda_+}{\lambda_\infty} \qquad ...(24)$$

Similarly $$t_- = \frac{\lambda_-}{\lambda_\infty} \qquad ...(25)$$

From Eqs. (24) and (25), it follows that the transport number of cation and anion can be determined if one knows the values of λ_+, λ_- and λ_∞.

In this case, oxidation takes place at the zinc electrode and reduction at the hydrogen electrode, as shown below:

Zinc electrode $Zn\,(s) \rightleftharpoons Zn^{2+} + 2e^-$

Hydrogen electrode $2H^+ + 2e^- \rightleftharpoons H_2(g)$

$\therefore$ *The net reaction is* $Zn(s) + 2H^+ \rightleftharpoons Zn^{2+} + H_2(g)\ (1\ atm)$

Nernst's Equation for Single Electrode Potential

From the Nernst theory it follows that the potential difference of the double layer formed at one electrode is termed as single electrode potential.

Let us consider a metal of valency 'n' dipping in a solution

$$M \rightleftharpoons M^{+n} + ne$$

Let P_1 be the osmotic pressure of its ions, P_3, be the solution pressure of the metal and E be the electrode potential, *i.e.* the difference of potential between the metal and solution. Now pass the current through the electrode reversible until 1 gm. ion of the metal gets dissolved. It means that the electricity required for the dissolution of 1 gm. ion of the metal will be nF* coulombs. Hence,

the electrical work done = NF*E volt-coulombs

i.e. $w_1 = nF^*E$ volt-coulombs

where F* is one faraday whose value is equal to 96500 coulombs and w_1 is the electrical work done .Go on diluting the solution till the osmotic pressure gets reduced from P_1 to $P_1 - dP_1$. At the same time the difference of potential between the metal and the solution is now changed from E to E – dE. At the potential difference E – dE, the electrical work done (w_2) to cause the dissolution of 1 gm. ion of the metal is given by

$$w_2 = (E - dE)\,nF^* \text{ volt-coulombs}$$

The difference in electrical energy of work = $w_1 - w_2$

or $$w_1 - w_2 = nEF^* - (E - dE)nF^*$$
$$= nF^*\,dE \text{ volt coulombs.}$$

The different in electrical energy is equal to the osmotic work done in transferring 1 gm. ion of the metal from P1 to P_1—dP_1 which is equal to the product of volume and change in pressure *i.e.*,

$$nF^*dE \rightleftharpoons VdP_1 \qquad ...(1)$$

where V is the volume of the solvent containing 1 gm. ion of the metal. For ideal solutions, osmotic pressures are directly proportional to the activities of ions, then $V = RT/P_1$. Thus, equation (2) reduces to

$$nF^*dE = \frac{RT\,dP_1}{P_1} \qquad ...(2)$$

$$nF^*\int dE = RT\int \frac{dP_1}{P_1}$$

$$nF^*E = 2.303\ RT\ \log P_1 + K \qquad ...(3)$$

where 'K' is the integration constant whose value will be evaluated as follows: when the solution pressure (P_2) is equal to osmotic pressure (P_1) of the ions, it means that no potential difference exists between the metal and the solution, *i.e.*, $E = 0$ and $P_1 = P_2$. Hence equation (3) becomes as

$$0 = 2.303\ RT\ \log P_2 + K$$

or

$$K = -2.303\ RT\ \log P_2 \qquad ...(4)$$

Substituting the value of K from eq. (4) in (3), we get

$$nF^*E = 2.303\ (RT\ \log P_1 - RT\ \log P_2)$$

$$E = 2.303\frac{RT}{nF^*}\log\frac{P_1}{P_2} \qquad ...(5)$$

Equation (5) is termed as the Nernst equation for electrode potential. Let us now consider tow similar electrodes dipping in two different solutions. In this case, the potential difference between the electrodes will be given by

$$E_1 - E_2 = \frac{2.303\ RT}{nF^*}\left(\log\frac{(P_1)_A}{P_2} - \log\frac{(P_1)_B}{P_2}\right)$$

$$= \frac{2.303\ RT}{nF^*}\log\frac{(P_1)_A}{(P_2)_B} \qquad ...(6)$$

where $(P_1)_A$ and $(P_1)_B$ denote the osmotic pressure of the metal ions in two solutions A and B respectively. As we know that the osmotic pressure is proportional to concentration, it means, that equation (6) may be written as

$$E_1 - E_2 = \frac{2.303\ RT}{nF}\log\frac{(c_1)_A}{(c_2)_B}$$

DERIVATION OF PHASE RULE

J. Willard Gibbs (1876) deduced a *mathematical* relation between the number of degrees of freedom (F), the number of components (C) and the number of phases (P) in a heterogeneous system at equilibrium. The relation is as follows :

$$F = C - P + 2$$

It was first derived by Gibbs from thermodynamic considerations and is known as *Gibbs phase rule.* Let us consider a system of 'C' components distributed between 'P' phases and let each phase contain all the 'C' components. To define the composition of each phase it is required to specify (C – l) concentration terms and hence the concentration of the remaining components will be defined by the difference. As there are 'P' phases the total number of variables for concentration alone is P (C – l).

As the process is being carried out at a constant temperature and pressure, it is required to add two more variables. Hence the total number of variables is becoming equal to P(C – 1) + 2.

We will now determine how many equations involving these variables are there when all the phases are in equilibrium. This can be shown in the following manner :

Consider a heterogeneous system of two phases (a and b) containing three components (1, 2, 3) in equilibrium. The chemical potential (μ) of these components in two phases can be represented as :

$$\mu_1\,(a),\ \mu_2\,(a),\ \mu_3\,(a),\ \mu_1\,(b),\ \mu_2\,(b),\ \mu_3\,(b)$$

As we are considering the closed system in equilibrium at a definite temperature, and pressure, the *Gibbs-Duhem* equation can be applied to it.

Thus, we get$\Sigma\mu\ dn = 0$.

Suppose a small amount δn_1 is transferred, under equilibrium conditions from phase a to b and therefore:

$$-\mu_1\,(a)\,dn_1 + \mu_1\,(b)\,dn_1 = 0$$

or $$\mu_1\,(a) = \mu_1\,(6)$$

This equation reveals that the equilibrium is between two phases when the chemical potential of a particular component is the same in each phase. This can be applied to all the components. It, therefore, follows that

$$\mu_2\,(a) = \mu_2\,(b);\ \mu_3\,(a) = \mu_3\,(b);\ \text{etc.}$$

Now, imagine the system is having three phases. Therefore, the chemical potential of a particular component must be same in all the phases, *i.e.*,

$$\mu_1(a) = \mu_1(b) = \mu_1(c)$$

Hence the number of independent equations defining the equilibrium between three phases are two only. In general, the above conditions also hold good for 'P' phases and 'C' components to give a set of following relations :

$$\mu_1(a) = \mu_1(\&) = \mu_1(c) = \ldots = \mu_1(P)$$
$$\mu_1(a) = \mu_2\,(b) = \mu_2(c) = \ldots = \mu_2\,(P)$$
$$\ldots \quad \ldots \quad \ldots \quad \ldots \quad \ldots$$
$$\ldots \quad \ldots \quad \ldots \quad \ldots \quad \ldots$$
$$\mu_c(a) = \mu_c(b) = \mu_c(c) = \ldots = \mu_c(P)$$

These equations constitute C(P – l) independent equations. Thus the degree of freedom F must be equal to

$$F = [P(C - 1) + 2] - [C(P - 1)]$$
$$= C - P + 2$$

This is the phase rule of Gibbs.

Discussion : If a component is missing from a particular phase, however, the number of concentration variables is decreased by one. But at the same time the number of possible equations is also decreased by one. Hence the value of (C–P), and therefore F. remains the same whether each constituent is present in every phase or not. This means that the phase rule is not restricted by the assumption made, and is generally valid under all conditions of distribution provided that equilibrium exists in the system.

Advantages of Phase Rule

(i) It gives a simple method of classifying equilibrium states of systems.

(ii) It confirms that the different systems having the same number of degrees of freedom behave in like manner.

(iii) It predicts the behaviour of systems when subjected to changes in the variables such as pressure, temperature and volume.

(iv) The phase rule is applicable to macroscopic systems. Therefore, it is not necessary to take into account about their molecular structures.

(v) It is applicable to physical as well as to chemical phase reactions.

(vi) Phase rule takes no account of nature of the reactants or products in phase reactions.

(vii) Phase rule predicts that a number of substances would remain in equilibrium or not in equilibrium if some of the substances have been transformed into the new substances.

Limitations of the Phase Role

(i) As the phase rule is applicable to heterogeneous systems in equilibrium, it is therefore of no use for such systems which are slow in reaching the equilibrium state.

(ii) As the phase rule is applicable to a single equilibrium state, it never tells about the number of other equilibrium possible in the system.

(iii) In phase rule various variables are temperature, pressure and composition. This phase rule does not consider the electric and magnetic influences. If such variables are considered, the factor 2 of the phase rule has to be adjusted accordingly.

(iv) All the phases in the system must be present under the same pressure, temperature and gravitational force.

(v) No liquid or solid phases should be finely divided otherwise their vapour pressures will differ from their normal values.

GIBBS EQUATIONS

The phase rule of Gibbs may be expressed in the simple form :

$$P + F = C + 2 \qquad ...(1)$$

where P is the number of *phases*, F is the number of *degrees of freedom* and C is the number of *components* in a system at equilibrium. Equation (1) does not involve the mass of phase as it has no influence on the state of equilibrium. In words, the phase rule (Eq. 1) may be stated as:

"In a heterogeneous system in equilibrium, the number of degrees of freedom plus the number of phases are equal to the number of components plus 2."

Before deriving the phase rule, it is necessary to explain the meaning of phase, component and degree of freedom. *Phase* : As stated earlier, a heterogeneous system consists of different parts of each of which is homogeneous in itself and is separated from others by bounding surfaces or interfaces. The different parts of heterogeneous system can be separated by mechanical methods. *These homogeneous, physically distinct and mechanical separable parts of the heterogeneous system in equilibrium are called phases. A phase may be gas, liquid or solid.*

Examples : (i) In a system consisting of ice, water and vapour, there are three phases; ice (solid phase), water (liquid phase) and vapour (gas phase). Each phase is homogeneous in itself and can be separated from the other phases by mechanical methods.

(ii) *All gases are completely miscible* : Therefore, a gaseous mixture irrespective of the number of gases present, consists of one phase only.

(iii) Two or more liquids which are miscible with one another constitute a single phase as there is no bounding surfaces separating the different liquids. Example is the alcohol and water which exist in one phase only as two are completely miscible. As there will be vapour in contact with the liquid, there will be two phases, *i.e.*, liquid and gas in miscible liquids.

(iv) A solid mixture consists of many as phases as are solids present. A heterogeneous equilibrium,

$$CaCO_3 \rightleftharpoons CaO + CO_2$$

involves three phases, solid ($CaCO_3$), solid (CaO) and gas (CO_2).

Thus, there are three phases, two solids and one gaseous.

Component : A component is an element or a compound present in a system. The concentration of this can undergo variation independently. But in the phase rule the component does not stand for the total number of constituents of the system. This term, is however, defined as :

"The number of components of a system at equilibrium is defined as the minimum number of molecular species in terms of which the composition of all the phases may be expressed quantitatively:

The meaning of this term will be made clear by considering some actual cases.

(i) Consider a water system consisting of the three phases :

$$\text{Ice} \rightleftharpoons \text{water} \rightleftharpoons \text{Vapour.}$$

Thus, water system is one component system as each phase is essentially a different physical form of the same chemical compound represented by H_2O.

(ii) Consider the dissociation of $CaCO_3$ by heat

$$CaCO_3 \rightleftharpoons CaO + CO_2$$

This equilibrium consists of three phases, two solids and one gas. At first sight, it seems that it is a three component system. This cannot be so as three constituents are not independent of each other as demanded by the definition. For example :

$$CaCO_3 = CaO + CO_2$$

or $$CaO = CaCO_3 - CO_2$$

or $$CO_2 = CaCO_3 - CaO.$$

Thus, the composition of all the three phases can be expressed by taking two of the constituents only. Hence it forms a two component system. For example :

(a) By taking CaO and CO_2 as the components

Phase Component

$$CaCO_3 = CaO + CO_2$$

$$CaO = CaO + 0\ CO_2$$

$$COa = 0\ CaO + CO_2$$

(b) By taking $CaCO_3$ and CO_2 as components

Phase Component

$$CaCO_3 = CaCO_3 + 0\ CO_2$$

$$CaO = CaCO_3 - CO_2$$

$$CO_2 = 0\ CaCO_3 + CO_2$$

(c) By taking $CaCO_3$ and CaO as components

Phase Component

$$CaCO_3 = CaCO_3 + 0\ CaO$$

$$CaO = 0\ CaCO_3 + CaO$$

$$CO_2 = CaCO_3 - CaO$$

(iii) Consider the dissociation of ammonium chloride in a closed vessel which is in equilibrium with its products, ammonia and hydrochloric acid

$$\underset{\text{solid}}{NH_4Cl} \rightleftharpoons \underset{\text{gas}}{NH_4Cl} \rightleftharpoons NH_3 + HCl$$

The system is considered as a one-component system because the composition of both the solid and gaseous phases can be stated in terms of NH_4Cl only. This is shown as below:

Phase	*Component*
Solid	NH_4Cl
Gaseous	$x.NH_3 + x.HC1$ or $x\ NH_4Cl$

If a small amount of NH_3 or HC1 is added to the equilibrium mixture then the number of components would become two. Suppose excess of hydrochloric acid is added, then the gaseous mixture will contain x moles of NH_3 + x moles of HCl + y moles of HCl. Tlherefore, the composition of two phases of ammonium chloride may be expressed as

Phase	*Component*
Solid NH_4Cl	NH_4Cl
Gas	.t.NHa +X.HCl –{ y.HCl
	$= x\ NH_4Cl + y\ HCl$

Thus, the minimum number of components by means of which the composition of both phases can be expressed arc the two components NH_4Cl and HCl. Thus, it is a two *component system*. Similarly it behaves as a two component system if ammonia is added to the equilibrium mixture.

Degree of freedom or variance of the system : The degree of freedom or variance of a system is defined as the least number of variable factors such as temperature, pressure and concentration which should be arbitrarily fixed in order to define the system completely.

(i) A gaseous mixture, say of carbon dioxide and nitrogen, is completely defined when three variables temperature, pressure and concentration are specified. Then, the system is said to have three degrees of freedom or to be trivariant.

(ii) If one consider a system consisting of unsaturated water vapour, two variables pressure and temperature are to be specified to define the system because pressure may have any value at a particular temperature. This system is said to have two degrees of freedom or bivariant.

PHASE REACTIONS

There are certain physical or chemical reactions which are accompanied by the appearance or disappearance of a phase. Such reactions are known as phase reactions.

A phase reaction is generally indicated in Fig. 2.7. The follows that the examples of phase reactions are as follows:

(i) Change of solid to liquid.

(ii) Change of liquid to solid.

(iii) Change of liquid to gas.

(iv) Change of solid to gas.

(v) Change of gas to liquid.

(vi) Change of gas to solid.

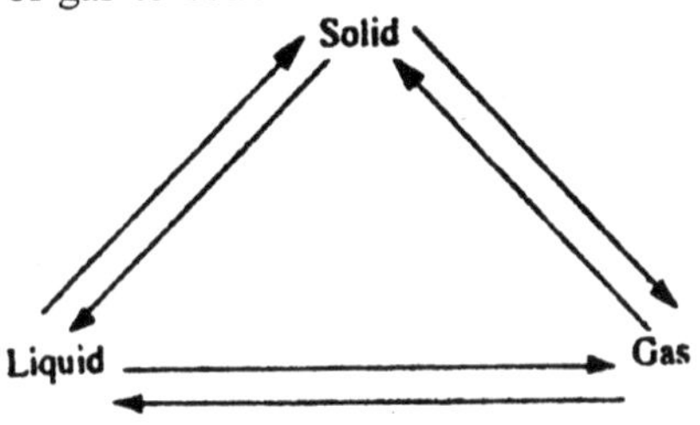

Fig. 2.7

From the above examples, it is evident that all processes of vaporisation, sublimation, freezing, melting, solidification, crystallisation, and precipitation are phase reactions.

All chemical processes may not be phase reactions, *e.g.*, the combination of gaseous ammonia and gaseous hydrogen chloride to form solid ammonium chloride is a phase reaction because the, reactants and products are in different phases. But the reaction between gaseous hydrogen and gaseous chlorine to form gaseous hydrogen chloride cannot be regarded as a phase reaction because all the, substances are in one phase.

CONDITIONS FOR EQUILIBRIUM BETWEEN PHASES

In order for each phase to be in equilibrium in a multiphase system, the following equilibrium conditions are to be fulfilled :

(a) *Thermal Equilibrium* : It means that the temperature of all the phases must be identical, otherwise there would occur the flow of heat from one phase to another. One can prove it thermodynamically as follows :

Suppose we consider two phases α and β at temperature T^α and T^β. Suppose S^α and S^β are the entropies of the two phases. Suppose the heat, dq, is transferred from phase α to β at equilibrium. Due to this heat transfer, the change of entropy of the system at equilibrium will be as follows :

$$dS = dS^\alpha + dS^\beta = 0 \qquad ...(1)$$

We know, $$dS^\alpha = -\frac{dq}{T\alpha} \text{ and } dS^\beta = \frac{dq}{T\beta} \qquad ...(2)$$

In equation (2), –ve and +ve signs are used to signify the heat loss and heat gain by the phases α and β respective ly.

On substituting equation (2) in (1), we get

$$-\frac{dq}{T\alpha} + \frac{dq}{T\beta} = 0$$

or $$T\alpha = T\beta \qquad ...(3)$$

From the equation (3), *it follows that the temperature of all the phases in equilibrium is same.*

(b) *Mechanical Equilibrium* : By this equilibrium it is meant that all the phases in equilibrium must be under the same pressure. It is not so, one phase would increase its volume at the expense of another. This may be deduced as follows :

Suppose there occurs an expansion in volume, dV, of α-phase at the expense of β-phase. Then at constant temperature, the change in Helmholtz free energy will become as follows :

$$dA = dA^\alpha + dA^\beta = 0 \text{ (At equilibrium)} \qquad ...(4)$$

we know $$dA^\alpha = P^\alpha\, dv \text{ and } dA^\beta = -P^\beta\, dv \qquad ..(5)$$

In equilibrium (5), positive sign indicates increase in volume of α-phase whereas negative sign indicates decrease in volume of β-phase.

On substituting equation (5) in (4), we get

$$P^{\alpha} = P^{\beta} \quad ...(6)$$

From equation (1), it follows *that the phases in equilibrium are under the same pressure.*

(c) *Chemical Equilibrium :* By this equilibrium, it is meant that the chemical potential of any components is same in all the phases at equilibrium.

Let us consider a closed system of P phases indicated by α, β, γ, ... having u total of C components indicated by 1, 2, 3, ... P in equilibrium and each at the same pressure and temperature. We know that the Gibbs tree energy (G) of each phase is a function of temperature, pressure and composition, *i.e.,*

$$G^{\alpha} = f(T, P, n_i)^{\alpha} \quad ...(7)$$

$$G^{\alpha} = f(T, P, n_i)^{\beta} \quad ...(8)$$

....

$$G^{P} = f\,(T, P, n_i)^{P} \quad ...(9)$$

where i = l, 2, 3, 4, ..., i, C The free energy change of the entire system will be the sum of change of free energies of each phase.

$$dG = dG^{\alpha} + + dG^{\beta} + dG^{v} + ...$$

For a multicomponent system, we know

$$dG = -S\ dT + V\ dp + \Sigma\mu_i\ dn$$

At constant temperature and pressure, dT = dP = 0 and therefore the above equation becomes as

$$(dG)r,\ p = \Sigma\mu_i\ dn_i$$

If there occurs a transfer of mass from one phase to another, we can write

$$dG = \sum_{i=1}^{0} \mu_i^{\alpha}\ dn_i^{\alpha} + \Sigma\mu_i^{\beta}\ dn_i^{\beta} + \Sigma\mu_i^{y}\ dn_i^{y} + ... \quad ...(10)$$

If we are considering a closed system at equilibrium (dG = 0), we can write

$$\Sigma\mu_i^{\alpha}\ dn_i^{\alpha} + \Sigma\mu_i^{\beta}\ dn_i^{\beta} + \Sigma\mu_i^{y}\ dn_i^{y} + ... = 0$$

At equilibrium total mass of each component is constant because we are considering a closed system. Therefore, we can write

$$\left.\begin{array}{l} dn_1^{\alpha} + dn_1^{\beta} + dn_1^{y} + ... + dn_1^{p} = 0 \\ dn_2^{\alpha} + dn_2^{\beta} + dn_2^{y} + ... + dn_2^{p} = 0 \\ ... \quad ... \quad ... \quad ... \quad ... \\ dn_i^{\alpha} + dn_i^{\beta} + dn_i^{y} + ... + dn_i^{p} = 0 \\ ... \quad ... \quad ... \quad ... \quad ... \\ dn_c^{\alpha} + dn_c^{\beta} + dn_c^{y} + ... + dn_c^{p} = 0 \end{array}\right\} \quad ...(11)$$

For all the equations in expression (11) to be zero for all possible variations of n, subject to the restrictions posed by equation (10), it is required that

$$\left.\begin{array}{l} \mu_1^{\alpha} = \mu_1^{\beta} = \mu_1^{y} = ... = \mu_1^{p} \\ \mu_2^{\alpha} = \mu_2^{\beta} = \mu_2^{y} = ... = \mu_2^{p} \\ ... \quad ... \quad ... \quad ... \\ \mu_c^{\alpha} = \mu_c^{\beta} = \mu_c^{y} = ... = \mu_c^{p} \end{array}\right\} \quad ...(12)$$

From expression (12), it is evident that for any component i in the system, the value of chemical potential (α_i) must be same in every phase provided the system is in equilibrium at constant T and P.

BUFFER SOLUTIONS OR BUFFERS

We know that the pH of neutral water is 7. If we add 1 cc of 1N HCl to 1 litre of water, the $[H^+]$ concentration in the solution becomes 10^{-3} gm ions and the pH of the solution becomes 3. Similarly, the addition of 1 cc of 1N NaOH in a litre of pure water changes the pH of the solution to 11.

Thus, we see that the pH of water changes by large amounts on the addition of small quantity of a strong acid or a strong base. But for certain biological and analytical purposes, we require solution whose pH should not change on keeping or on addition of small amounts of a strong acid or a strong base.

Such solutions which oppose the change in their pH on the addition of small amount of an acid or a base are called buffer solutions. Some characteristics of buffer solution are:

(i) Their pH should not change on keeping.

(ii) Their pH should remain same on dilution.

(iii) Their pH should remain same on the addition of small amounts of strong acid and base.

Action of Buffers

A buffer solution invariably consists of a weak acid and its salt solution (acidic buffer) or a weak base and its salt solution (basic buffer). Some examples of buffer solutions are:

Constituents	*pH Range*
Glycine + Glycine hydrochloride	1.0 — 3.7
Phthalic acid + Potassium hydrogen phthalate	2.2 — 3.8
Acetic acid + Sodium acetate	3.7 — 5.6
Disodium citrate + Trisodium citrate	5.0 — 6.3
Monosodium phosphate + Disodium phosphate	5.8 — 8.0
Boric acid + Borax	6.8 — 9.2
Borax + Sodium hydroxide	9.2 — 11.2
Disodium phosphate + Trisodium phosphate	11.0 — 12.0

Action of Acidic Buffer

Let us illustrate its action by considering a mixture sodium acetate and acetic acid in water.

$$CH_3COOH \rightleftharpoons CH_3COO^- + H^+ \text{ (feebly ionised)}$$

$$CH_3COONa \rightleftharpoons CH_3COO– + Na^+ \text{ (highly ionised)}$$

As CH_3COONa is almost fully ionised, the acetate ions so produced suppress the ionisation of the acetic acid, so that the mixture contains unionised acetic acid molecules and a large number of acetate ions. When a few drops of an acid, say HCl, are added the H^+ ions from the acid added react with the CH_3COO^- ions to form unionised or feebly ionised acetic acid:

$$CH_3COO^- + H^+ \rightarrow CH_3COOH$$

On the other hand, when a few drops of an alkali, say, NaOH are added, the OH^- ions react with CH_3COOH to form unionised H_2O.

$$CH_3COOH + OH^- \rightarrow CH_3COO^- + H_2O$$

Thus, the addition of small amounts of the acid or alkali to a buffer solution does not alter the H^+ ions concentration of the buffer. In other words, the pH of the buffer remains unchanged.

Action of a Basic-buffer

Let us illustrate the action of a basic buffer by considering a mixture of NH_4OH and NH_4Cl in water. They exhibit the following equilibria

$$NH_4OH \rightleftharpoons NH_4^+ + OH^- \text{ (weakly ionised)}$$

$$NH_4Cl \rightleftharpoons NH_4^+ + Cl^- \text{ (completely ionised).}$$

The presence of the excess of NH_4^+ ions from completely ionised NH_4Cl suppresses the ionisation of NH_4OH, so that the mixture solution contains unionised NH_4OH molecules and a large number of NH_4^+ ions. When a few drops of alkali is added to this buffer solution, the OH^- ions produced from the alkali react with NH_4^+ present in solution to form unionised NH_4OH.

$$NH_4^+ + OH^- \rightarrow NH_4OH$$

On adding a few drops of the acid say HCl, the H^+ ions provided by the acid react with NH_4OH to give unionised H_2O.

$$NH_4OH + H^+ \rightarrow NH_4^+ + H_2O$$

Thus, the addition of small amounts of the alkali or acid to the above solution does not alter its H^+ ion concentration. In other words, the pH of buffer remains unchanged.

Mathematical Expression for the pH of an Acidic Buffer

Consider a solution having a weak acid HA and its salt BA, with a strong base. The ionisation of the weak acid is represented as:

$$HA \rightleftharpoons H^+ + A^-$$

or

$$K_a = \frac{[H^+][A^-]}{[HA]} \quad \text{or} \quad [H^+] = K_a \frac{[HA]}{[A^-]} \qquad ...(1)$$

where K_a is the dissociation constant of the weak acid., The salt BA is almost completely ionised. Thus, high concentration of the A^- ions provided by almost complete ionisation of the salt BA suppresses the ionisation of the weak acid, so that it may be assumed that the free A^- ions are entirely due to the salt. Thus

$$[A^-] = [\text{Salt}] \text{ and } [HA] = [\text{Acid}] \qquad ...(2)$$

Substituting Eq. (2) in (1), we get

$$[H^+] = K_a \frac{[\text{Acid}]}{[\text{Salt}]}$$

Taking logarithm of both sides, we get

$$\log [H^+] = \text{loh } K_a + \log \frac{[\text{Acid}]}{[\text{Salt}]}$$

or
$$-\log [H^+] = -\text{loh } K_a - \log \frac{[\text{Acid}]}{[\text{Salt}]}$$

or
$$pH = pK_a - \log \frac{[\text{Acid}]}{[\text{Salt}]} \quad ...(3)$$

$$= PK_a + \log \frac{[\text{Salt}]}{[\text{Acid}]}$$

where $pK_a = -\log K_a$. Equation (3) is known as Henderson-Hasselbalch equation.

Mathematical Expression for the pH of a Basic Buffer

Consider a weak base BOH and its salt AB. The dissociation of the weak base may be written as

$$BOH \rightarrow B^+ + OH^-$$

or
$$K_a = \frac{[B^+][OH^-]}{[BOH]} \text{ or } [OH^-] = \frac{K_b[BOH]}{[B^+]} \quad ...(1)$$

where K_b is the dissociation constant of the weak base.

The high concentration of B^+ ions produced by the complete ionisation of the salt, which thus suppresses the ionisation of the weak base so that it may be assumed that free B^+ ions in the solution are entirely due to the salt. Thus,

$$[B^+] = \text{Salt and } [BOH] = \text{Base}$$

Thus, equation (1) may be put as

$$[OH^-] = K_b \frac{[\text{Base}]}{[\text{Salt}]}$$

Taking the logarithm of the above equation, we obtain

$$\log [OH^-] = \log K_b + \log [\text{Base}] - \log [\text{Salt}]$$

or
$$-\log [OH^-] = -\log K_b - \log [\text{Base}] + \log [\text{Salt}]$$

or
$$pOH = pK_b + \log \frac{[\text{Salt}]}{[\text{Base}]} \quad ...(2)$$

But pH = 14 – pOH, Equation (2) becomes as

$$pH = 14 - \left(pK_a + \log\frac{[Salt]}{[Base]}\right) \quad ...(3)$$

Now $[H^+][OH^-] = k_w = 10^{-14}$ or $\log k_w = -\log 10 = -14$

or $$14 = -\log k_w \quad ...(4)$$

From Eqs. (3) and (4), we get

$$pH = -\log k_w - \left(pK_b + \log\frac{[Salt]}{[Base]}\right)$$

$$= -\log k_w - \left(-\log K_b + \log\frac{[Salt]}{[Base]}\right)$$

$$= -\log\frac{k_w}{k_b} + \log\frac{[Base]}{[Salt]} \quad ...(5)$$

HYDROLYSIS

Pure water is neutral and is feebly ionised into H^+ and OH^- ions as:

$$H_2O \rightarrow H^+ + OH^-$$

The concentration of H^+ and OH^- ions is same. Now when a salt is dissolved in water, it ionises into ions, *e.g.*, the salt AB will ionise as:

$$AB \rightarrow A^+ + B^-$$

The ions of the salt may react with the ions of water to give

$$A^+ + OH^- \rightarrow AOH$$

$$B^- + H^+ \rightarrow HB$$

Now, the solution will be acidic or alkaline depending upon the relative strengths of the acid HB and the base AOH. If the acid is strong and the base is weak, then the solution will be acidic in nature. If the base is strong and the acid is weak, the solution will be basic in nature. Thus, this phenomenon in which a salt interacts with water to produce either an acidic or alkaline solution is termed as hydrolysis. Actually, hydrolysis involves the interaction of the ions of a salt with the water to produce H^+ or OH^- ions in solution. Thus, it may be defined as:

"*The interaction between the ions of salt and the ions of water to given free H^+ or OH^- ions in solution.*"

Example of Hydrolysis

Salt may be divided into four categories:

(a) Salts of Strong Acid and Strong Bases

These salt, when dissolved in water, produce a strong acid and a strong base, so that the relative proportions of H^+ and OH^- ions in solution remain unchanged and the solution remains neutral. IN simple words, it can be said that the salts of strong acids and bases do not undergo hydrolysis. Let us consider the dissolution of sodium chloride in water.

$$NaCl \rightarrow Na^+ + Cl^-$$

$$H_2O \rightarrow OH^- + H^+$$

$$Na^+ + OH^- \rightarrow NaOH \text{ (highly ionised)}$$

$$H^+ + Cl^- \rightarrow HCl \text{ (highly ionised)}$$

But the acid, HCl and the base, NaOH being equally strong are almost completely ionised. Thus, the concentration of H^+ and OH^- is not altered and the solution remains neutral. Examples of such salts are NaCl, KCl, Na_2SO_4, K_2SO_4, etc.

(b) Salts of Weak Acids and Strong Bases

These salts (like Na_2CO_3; CH_3COONa, KCN, etc.), when dissolved in water undergo hydrolysis in water to give an alkaline solution. For example,

$$CH_3COONa \rightarrow CH_3COO^- + Na^+$$

$$H_2O \rightarrow H^+ + OH^-$$

$$CH_3COO^- + H^+ \rightarrow CH_3COOH$$

$$Na^+ + OH^- \rightarrow NaOH$$

NaOH being a strong base is almost completely ionised and the concentration of OH^- ions in solutions is not altered. However, CH_3COOH being a weak acid is feebly ionised. Thus, the concentration of H^+ ions in solution falls. Thus, the solution has more of OH^- ions than the H^+ ions. Hence, it is alkaline in nature.

(c) Salts of Strong Acids and Weak Bases

Examples are NH_4Cl, $CuSO_4$, etc. When such salts are dissolved in water, they produce acidic solutions. Consider the dissolving of ammonium chloride.

$$NH_4Cl \rightarrow NH_4^+ + Cl^-$$
$$H_2O \rightarrow H^+ + OH^-$$
$$NH_4^+ + OH^- \rightarrow NH_4OH$$
$$H^+ + Cl^- \rightarrow HCl$$

HCl being a strong acid is almost completely ionised and the concentration of the H^+ ions in solution is not affected. NH_4OH, however, is a weak base and feebly ionises to give only few OH^- ions in solution. Thus, the solution has a large concentration of the H^+ ions that the OH^- ions. Hence the solution is acidic in nature.

(d) Salts of Weak Acid and Weak Bases

Examples are CH_3COONH_4, $(NH_4)_2CO_3$, etc. Such salts when dissolved in water yield acidic or basic or even neutral solutions, depending upon the nature of the acid and the base produced. Consider the dissolving of ammonium acetate in water.

$$CH_3COONH_4 \rightarrow CH_3COO^- + NH_4^+$$
$$H_2O \rightarrow H^+ + OH^-$$
$$CH_3COO^- + H^+ \rightarrow CH_3COOH$$
$$NH_4^+ \; OH^- \rightarrow NH_4OH$$

Both CH_3COOH and NH_4OH are weak and are equally but feebly ionised. Hence the solution contains practically the same number of the H^+ and OH^- ions. Therefore, the solution is almost neutral. In case of $(NH_4)_2CO_3$, however, the solution is slightly basic because NH_4OH is relatively a stronger base than the H_2CO_3 acid.

MATHEMATICAL TREATMENT OF HYDROLYSIS

(a) Salt of Strong Acids and Weak Bases

(i) *Hydrolysis constant.* The hydrolysis of a salt of a strong acid and a weak base may be represented as:

$$A^+B^+ + H_2O \rightarrow HOH + B^- + H^+$$

or

$$A^+ + H_2O \rightarrow AOH + H^+$$

Applying the law of mass action to the above equilibrium,

$$K = \frac{[AOH][H^+]}{[A^+][H_2O]}$$

As water is taken in large excess, it means that its active mass remains practically constant. Therefore,

$$K[H_2O] = \frac{[AOH][H^+]}{A^+}$$

$$K_h = \frac{[AOH][H^+]}{A^+}, \text{ where } K_h = K[H_2O] \quad ...(1)$$

The constant K_h is called the hydrolysis constant of the salt.

(ii) *Relation between K_h, K_b and K_w.* The dissociation of the weak base produced in the hydrolysis of a strong acid and a weak base may be represented as:

$$AOH \rightarrow A^+ + OH^-$$

Applying the law of mass action,

$$K_b = \frac{[A^+][OH^-]}{[AOH]} \quad ...(2)$$

where K_b is the dissociation constant of the weak base.

Multiplying equations (1) and (2), we obtain

$$[H^+][OH^-] = K_h \times K_b$$

But $[H^+][OH^-] = K_w$, the ionic product of water

$\because$ $K_w = K_h \times K_b$

or

$$K_h = \frac{k_w}{K_b} \quad ...(3)$$

From equation (3), it is evident that the *weaker the base, the greater will be the hydrolysis constant of the salt.*

(iii) *Degree of hydrolysis:* Degree of hydrolysis of salt may be defined as:

"The fraction of the total salt hydrolysed when the equilibrium has been established is called the degree of hydrolysis."

It is denoted by the letter '*h*'. In simple words, the extent to which the salt has been hydrolysed is known as the degree of hydrolysis. If *c* is the initial concentration of the salt in moles per litre and *h* is the degree of hydrolysis on the attainment of equilibrium, then the concentrations of various species will be represented as:

$$\underset{c(1-h)}{A^+} + \underset{\text{Excess}}{H_2O} \rightarrow \underset{ch}{AOH} + \underset{ch}{H^+} \quad ...(4)$$

Applying the law of mass action,

$$K_h = \frac{(ch)(ch)}{c(1-h)} \qquad ...(5)$$

$$= \frac{ch^2}{(1-h)} = \frac{ch^2}{(1-h)}$$

If the base is very weak, the value of h is very small as compared to unity. It means that 1 – h in the equation (5) may be replaced by 1, *i.e.*,

$$K_h = ch^2$$

or

$$h = \sqrt{\left(\frac{K_h}{c}\right)} \qquad ...(6)$$

From equation (3), we know

$$K_h = \frac{k_w}{K_c} \qquad ...(7)$$

Replacing the above value of K_h in equation (6), we obtain

$$h = \sqrt{\left(\frac{k_w}{cK_b}\right)} \qquad ...(8)$$

From the relation (8), it follows that the degree of hydrolysis of a salt of a weak base and a strong acid is:

(i) *directly proportional to the ionic product of water*

(ii) *inversely proportional to the value of K_b, and*

(iii) *inversely proportional to the value of c.*

From the above it follows that degree of hydrolysis increases if the base is weaker *i.e.*, if K_b is smaller, and also increases if the concentration decreases or the dilution increases.

As k_w increases much more than K_b, it means that the degree of hydrolysis at a given concentration increases with the increase in temperature.

(iv) Relation between pH, pK_b and c. From equation (4), it follows:

$$[H^+] = ch$$

$$= c\sqrt{\left(\frac{k_w}{cK_b}\right)} \qquad \text{[From equation (8)]}$$

$$= \sqrt{\left(\frac{ck_w}{K_b}\right)} \qquad ...(9)$$

Taking logarithms and changing the signs throughout, Eq. (9) becomes as:

$$-\log[H^+] = -\frac{1}{2}\log k_w + \frac{1}{2}\log K_b - \frac{1}{2}\log c$$

$$pH = \frac{1}{2}pk_w - pK_b - \frac{1}{2}\log c \qquad ...(10)$$

From equation (10), it follows that the pH of the solution must be less than $\frac{1}{2}pK_w$ and is less than 7.0 and so solutions of salts of the type under consideration will be acid solutions.

(b) Salts of Weak Acids and Strong Bases

(i) *Hydrolysis constant*: The hydrolysis of a salt of a weak acid and a strong base may be represented as:

$$(A^+ + B^-) + H_2O \rightarrow HB + A^+ + OH^-$$

$$B^- + H_2O \rightarrow HB + OH^-$$

Applying the law of mass action,

$$K = \frac{[OH^-][HB]}{[B^-][H_2O]} \qquad ...(11)$$

As water is taken in large excess, its active mass remains practically constant. Therefore,

$$K[H_2O] = \frac{[OH^-][HB]}{[B^-]}$$

or

$$K_h = \frac{[OH^-][HB]}{[B]} \qquad ...(12)$$

where K_h is known as the hydrolysis constant of the acid.

(ii) *Relation between K_h, K_a and k_w*: The dissociation of weak acid, HB, produced during hydrolysis may be represented as:

$$HB \rightleftharpoons H^+ + B^-$$

Applying the law of mass action,

$$K_a = \frac{[H^+][B^-]}{[HB^-]} \qquad ...(13)$$

where K_a is known as the dissociation constant of the weak acid.,

Multiplying Eq. (12) by (13), we get

$$[H^+][OH^-] = K_a \times K_h$$

But $\quad [H^+][OH^-] = k_w$

$\therefore \quad k_w = K_a \times K_b$

or $$K_b = \frac{k_w}{K_a} \qquad ...(14)$$

From eq. (14), it follows that the hydrolysis constant, K_h, of the salt varies inversely as the dissociation constant, K_a, of the weak acid. Therefore, the weaker the acid, the greater is the hydrolysis constant of the salt.

(iii) *Degree of hydrolysis:* If c is the initial concentration fo the salt in moles per litre and h is the degree of hydrolysis on the attainment of equilibrium, the concentration of various species may be represented as:

$$\underset{c(1-h)}{B^-} + \underset{\text{excess}}{H_2O} \rightarrow \underset{ch}{HB} + \underset{ch}{OH^-} \qquad ...(14A)$$

Applying the law of mass action,

$$K_h = \frac{(ch)(ch)}{c(1-h)} = \frac{ch^3}{1-h} \qquad ...(15)$$

If the acid is very weak, h is negligible as compared to unity, *i.e.*, $1 - h \approx 1$. Thus, equation (15) becomes as:

$$K_h = ch^2$$

or $$h = \sqrt{\left(\frac{K_h}{c}\right)}$$

From equation (14), we have

$$K_h = \frac{k_w}{K_a}$$

$$\therefore \quad h = \sqrt{\left(\frac{k_w}{cK_a}\right)} \qquad ...(16)$$

From equation (16), it follows that weaker the acid, greater is h, *i.e.*, the degree of hydrolysis. As value of k_w is also increased by a rise in temp., it means that the degree of hydrolysis 'h' will considerably increase with rise of temperature It is also seen that degree of hydrolysis 'h' also increases if the concentration decreases *i.e.*, dilution increases.

(iv) *Relation between pH, pK_a and c:* From Eq. (14A), the concentration of $[OH^-]$ ion is given by

$$[OH^-] = ch$$

But $$[H^+][OH^-] = k_w$$

$$[H^+] = \frac{k_w}{ch} \quad ...(17)$$

From equation (16), we have

$$h = \sqrt{\left(\frac{k_w}{cK_a}\right)}$$

Substituting the value of h in (17), we have

$$[H^+] = \frac{k_w}{c}\sqrt{\left(\frac{cK_a}{k_w}\right)} = \sqrt{\left(\frac{k_w K_a}{c}\right)} \quad ...(18)$$

Taking logarithms and changing the signs throughout, equation (18) becomes as

$$-\log[H^+] = -\frac{1}{2}\log k_w - \frac{1}{2}\log K_a + \frac{1}{2}\log c$$

or $$pH = pk_w + \frac{1}{2}pK_e + \frac{1}{2}\log c \quad ...(19)$$

From equation (19), it follows that pH or alkalinity of a solution of the salt of a weak acid and strong base increases with decreasing and strength, *i.e.*, increasing pK_a and increasing concentration.

(c) Salt of Weak Acids and Weak Bases

(i) *Hydrolysis constant*: The hydrolysis of the salt of a weak acid and a weak base may be represented as:

$$AB + H_2O \rightleftharpoons AOH + HB$$

Ionically, the above equilibria can be represented as:

$$A^+ + B^- + H_2O \rightleftharpoons \underset{\text{weak base}}{AOH} + \underset{\text{weak acid}}{HB} \quad ...(19A)$$

Applying the law of mass action

$$K = \frac{[AOH][HB]}{[A^+][B^-][H_2O]}$$

$$K[H_2O] = \frac{[AOH][HB]}{[A^+][B^-]}$$

$$K_h = \frac{[AOH][HB]}{[A^+][B^-]} \quad ...(20)$$

where K_h is the hydrolysis constant of salt.

(ii) Relation between K_h, k_w, K_a, and K_b: The dissociation of the weak acid and weak base may be represented as:

$$HB \rightarrow H^+ + B^- \quad ...(21)$$

and

$$AOH \rightarrow A^+ + OH^- \quad ...(22)$$

Applying the law of mass action to the above equilibria (21) and (22),

$$K_a = \frac{[H^+][B^-]}{[HB]} \quad ...(23)$$

$$K_b = \frac{[A^+][OH^-]}{[AOH]} \quad ...(24)$$

where K_a and K_b are the dissociation constants of the weak acid and base respectively. Multiplying Eqs. (20), (23) and (24),

$$[H^+][OH^-] = K_a \times K_b \times K_h$$

But $$[H^+][OH^-] = k_w$$

$\therefore$ $$K_w = K_a \times K_b \times K_h$$

or $$K_h = \frac{k_w}{K_a \times K_b} \quad ...(25)$$

(iii) *Degree of hydrolysis:* If c is the initial concentration of the salt in moles/litre and 'h' is the degree of hydrolysis, then the concentration of various species of equilibria (19A) may be represented as:

$$\underset{c(1-h)}{A^+} + \underset{c(1-h)\text{ excess}}{B^- + H_2O} \rightarrow \underset{ch}{AOH} + \underset{ch}{HB} \quad ...(26)$$

Applying the law of mass action,

$$K_h = \frac{[ch][ch]}{c(1-h)c(1-u)} = \frac{h^2}{(1-h)^2} \quad ...(27)$$

If h being very small, h may be neglected and 1 – h may be taken as unity. Thus, equation (27) becomes as:

$$K_h = h^2 \qquad ...(28)$$

From Eq. (28) it follows that degree of hydrolysis of a weak acid and a weak base is independent of the dilution or concentration.

(iv) *Relation between pH, pK_w, pK_a and pK_b;* From equation (23), it follows that

$$[H^+] = K_a \frac{[HB]}{[B^-]} \qquad ...(29)$$

From equation (26), we get

$$B^- = c(1 - h) \text{ and } HB = ch \qquad ...(30)$$

Substituting equation (30) in (29), we have

$$[H^+] = K_a \frac{ch}{c(1-h)} = K_a \frac{h}{1-h}$$

$$= K_a(K_h)^{1/2} \qquad \text{[from equation (27)]}$$

$$= K_a\left(\frac{k_a}{K_a K_b}\right)^{1/2} \qquad \text{[from equation (25)]}$$

$$= \left(\frac{k_w K_a}{K_b}\right)^{1/2}$$

$$= K_4(K_h)^{1/2} \qquad \text{[from equation (27)]}$$

$$K_a = \left(\frac{K_w}{K_b}\right)^{1/2} \qquad \text{[from equation (25)]}$$

$$= \left(\frac{k_w K_a}{K_a K_b}\right) \qquad ...(31)$$

Taking logarithm of both sides obtain, we get

$$-\log[H^+] = -\frac{1}{2}\log k_w - \frac{1}{2}\log K_a + \frac{1}{2}\log K_b$$

$$pH = \frac{1}{2}pk_a + \frac{1}{2}pK_a - \frac{1}{2}pK_b \qquad ...(32)$$

Now three different cases may arise:

(i) When $K_a = K_b$, it follows from equation (32), $pH = \frac{1}{2}pk_w = 7$, *i.e.*, the solution will be neutral.

(ii) When $K_a > K_b$; $pH < 7$, *i.e.*, the solution will be acidic.

(iii) When $K_a < K_b$; pH > 7, *i.e.*, the solution will be alkaline.

Determination of Degree of Hydrolysis

(a) From Colligative Properties

The colligative property of a solution number of the solute particles. Consider, 1 gm mole of the salt of weak acid and strong base like CH_3COONa which gets hydrolysed in the solution as:

$$\underset{1-x}{CH_3COO^-} + \underset{1-x}{Na^+} + H_2O \rightarrow \underset{x}{CH_2COOH} + \underset{x}{Na^+} + \underset{x}{OH^-}$$

where x is the degree of hydrolysis. Then,

total number of particles before hydrolysis = 1 + 1 = 2

and the total number of particles after hydrolysis

$$= 1 - x + 1 - x + x + x + x = 2 + x$$

$$\because \quad \frac{\text{Observed freezing point depression}}{\text{Calculated freezing point depression}} = \frac{2+x}{2}$$

From the above relation, x can be calculated.

(b) From Conductance Measurements

The method is based upon the simple assumption that the equivalent conductance of an aqueous solution of a salt of a weak acid and strong base (or a weak base and strong acid) is due to the sum of equivalent conductances of strong base (or a weak base and strong acid) is due to the sum of equivalent conductances of

(a) the unhydrolysed salt and

(b) the free OH^- (or the H^+) ions produced by hydrolysis.

The weak acid or weak base formed during hydrolysis do not contribute anything towards the equivalent conductance of the salt solution at infinite dilution.

Procedure: The procedure involves the following steps:

(i) An aqueous solution for salt of known dilution is prepared. Now determine the equivalent conductance at a certain dilution, say, V, *i.e.* l_V.

(ii) Now mix the aqueous solution of the salt with excess of the weak acid or weak base as the case may be or the same which is produced during hydrolysis. This addition of weak acid or weak base checks the hydrolysis of the salt. Now determine the equivalent conductance of the mixture at the same dilution as before. Let it be λ'_V. Actually, this is the equivalent conductance of the unhydrolysed salt.

(iii) The equivalent conductance of the strong base or acid at infinite dilution is also determined by Kohlrausch's law. Let it be λ_∞.

Calculation: From the above values of the various equivalent conductances, the degree of hydrolysis may be calculated as follows:

Consider a salt AB of a strong acid and a weak base. Its hydrolysis may be represented as:

$$A^+B^- + H_2O \rightleftharpoons AOH + H^+ + B^-$$

or

$$\underset{1-h}{A^+} + H_2O \rightleftharpoons \underset{h}{AOH} + \underset{h}{A^+}$$

If h is the degree of hydrolysis, then

Concentration f unhydrolysed salt = 1 – h

Concentration of free H^+ ions = h

Equivalent conductance due to unhydrolysed salt = $\lambda'_V - (1-h)$ and equivalent conductance due to H^+ ions = $\lambda_\infty \times h$

∴ Equivalent conductance of hydrolysed salt,

$$\lambda_V = \lambda_\infty \times h + \lambda'_V (1-h)$$

or

$$h = \frac{\lambda_V - \lambda'_V}{\lambda_\infty - \lambda'_V}$$

This is known as Bredig's method.

(c) pH Determination Method

In case of the hydrolysis of a salt of a strong acid and a weak base,

$$A^+ + H_2O \rightleftharpoons AOH + H^+$$

and the value of $[H^+]$ is given as $[H^+] = ch$ ...(1)

where 'C' is the concentration of the salt in gm moles/litre and h is the degree of hydrolysis.

Taking logarithm of both sides of equation (1), we obtain

$$\log [H^+] = \log ch$$

or

$$-\log [H^+] = -\log ch$$

or $$pH = -\log ch \qquad ...(2)$$

In actual practice the aqueous solution of the salt is taken and its pH is determined with the help of pH meter. Thus, one can calculate the degree of hydrolysis.

(d) From the Dissociation Constants of the Weak Acids or Weak Bases

As we know in the case of

(i) a salt of a weak acid and a strong base

$$h = \sqrt{\left(\frac{k_w}{K_a \times C}\right)}] \qquad ...(1)$$

(ii) a salt of a weak base and a strong acid

$$h = \sqrt{\left(\frac{k_w}{K_b \times C}\right)} \qquad ...(2)$$

(iii) a salt of a weak acid and a weak base

$$h = \sqrt{\left(\frac{k_w}{K_a K_b}\right)} \qquad ...(3)$$

Knowing dissociation constant of acid or base (K_a or K_b) or both as the case may be, it is a simple matter to calculate the degree of hydrolysis of a given salt at a given concentration.

(e) Distribution Method

This method can also be used provided the free acid or base formed during hydrolysis is soluble in a immiscible solvent like benzene. Let us illustrate this principle by considering the hydrolysis of aniline hydrochloride. When aniline hydrochloride is hydrolysed, the following equilibria are established in solution.

$$C_6H_5NH_3^+Cl^- + H_2O \rightarrow C_6H_5NH_3^+OH^- + HCl$$

$$\downarrow$$

$$C_6H_5NH_2 + H_2O$$

Applying the law of mass action, the hydrolysis constant K_b is given as:

$$K_h = \frac{[\text{Concentration of free base}]\,[\text{Concentration of free acid}]}{[\text{Concentration of unhydrolysed salt}]}$$

If c is the initial concentration of the salt in moles per litre, then the degree of hydrolysis is given by:

$$h = \left(\frac{K_h}{c}\right)^{1/2} \qquad ...(1)$$

The concentration of aniline formed by hydrolysis of aniline hydrochloride can be determined as follows:

(i) A known amount of aniline is distributed between benzene and water. The concentration of aniline in the benzene layer is measured by passing dry HCl gas through it and weighing the amount of aniline 'hydrochloride precipitated. Difference from the total weight of aniline taken gives the concentration of aniline in water layer. When distribution law is applied, distribution coefficient 'K' is given by

$$K = \frac{\text{Aniline in benzene}}{\text{Aniline in water}}$$

(ii) Now shake a known weight of aniline hydrochloride with a known weight of benzene and water. The free aniline formed distributes itself between benzene and water. Amount of the aniline in the benzene layer is determined by precipitating as aniline hydrochloride by passing dry HCl gas.

Then, calculate the amount of aniline in the water layer by using the distribution coefficient determined in the first step.

The sum of the weights of aniline in the benzene and water layers gives the total weights of aniline formed by hydrolysis. The concentration of free HCl (in moles per litre) will be the same as of aniline in moles litre.

The concentration of unhydrolysed salt (c)

= [Initial concentration of the salt taken]

– [the loss of concentration due to hydrolysis]

= [Initial concentration of the salt]

– [Concentration of aniline formed by hyrolysis]

Substituting the various values in (1), the hydrolysis constant or the degree of hydrolysis of aniline hydrochloride can be determined.

3

Solids and Their Properties

INTRODUCTION

There is a close relationship between the physical properties of a solid and its structure and chemical composition. We shall now examine a few of the important properties of solids to illustrate the wide variety of possibilities. Some of these properties are exploited for various new innovations in electronic and magnetic devices such as transistors, computers, telephones, etc. The information explosion in inorganic and organic solid state materials has been so great in recent years that it would be difficult to list representative examples of all classes of materials in an elementary chemistry text of this sort.

Magnetic Properties

Materials can be divided into different classes depending on their response tq, magnetic fields. Diamagnetic materials (like NaCI or ZrO_2) are weakly repelled by magnetic fields. In paramagnetic materials there are permanent magnetic dipoles due to the presence of atoms, ions or molecules with unpaired electrons (*e.g.*, O_2, NO, Na atoms, Ti_2O_3, VO_2) and these materials are attracted by magnetic fields; they, however, lose their magnetism in the absence of a magnetic field. Solids like TiO_2 which should be diamagnetic often exhibit paramagnetism due to the presence of slight non-stoichiometry. Unlike paramagnetic materials, ferromagnetic materials show permanent magnetism even after the magnetic field is removed (*e.g.*, Fe, Cu_2Mn, Sn, EuO, CrO_2). The explanation for the large magnetization in these materials is that there are domains of magnetization (consisting of about a million atoms or ions), all of which cooperatively direct their magnetic moments (spans) in the same direction.

Once such a material is magnetized (with all domains oriented), it remains permanently that way. Iron, cobalt, and nickel are the only three elements which show ferromagnetism at room temperature. Manganese has most of the properties required for ferromagnetism, but the metal atoms are too close; however, addition of copper to manganese enhances the spacing and the alloy is ferromagnetic. We have just learnt that a spontaneous alignment of magnetic moments in the same direction gives rise to ferromagnetism. If the alignment of moments is in a compensatory way so as to give zero net moment, then we get antiferromagnetism in the material (*e.g.*, MnO).

Ferrimagnetism is found when the moments are aligned in parallel and antiparallel directions in unequal numbers resulting in a net moment (*e.g.*, Fe_3O_4, ferrites of the formula $M^{+2}Fe_2O_4$ (M = Mg; Cu, 3n, etc.). We have shown the spin alignments in antiferromagnetic MnO and ferrimagnetic Fe_3O_4. All these magnetically ordered solids transform to the paramagnetic state at a higher temperature due to the randomization of spins.

Table 1 : Magnetic properties of typical transition metal oxides.

TiO(p)	VO(p)	CrO(?)	MnO(af)	FeO(af)	CoO(af)	NiO(af)CuO(p)
Ti_2O_3(p)	V_2O_3(af)	Cr_2O_3(af)	Mn_2O_3(af)	Fe_2O_3(afw)	CO_3O_4(fe)	–Cu_2O(d)
TiO_2(d)	VO_2(p)		MnO_2(af)	Fe_3O_4(af)		
	V_2O_5(d)		Mn_3O_4(p)			

p = paramagnetic; ? = not known; af = antiferromagnetic; afw = antiferromagnetic with parasitic ferromagnetism; fe = temmagnetic; f = ferromagnetic.

For example V_2O_3 and NiO transform from the anti-ferromagnetic phase to the paramagnetic phase at 150°K and 523°K respectively: Ferromagnetic CrO_2 becomes paramagnetic at 320°K; a similar transition in ferrimagnetic Fe_3O_4 is at 850°K. Magnetic properties of a few transitional metal oxides.

Dielectric Properties

In insulators, electrons are closely bound to individual atoms or ions and they do not generally migrate under an applied electric field. However, dipoles are created by a shift in the charge resulting in polarization.

(i) These dipoles may align themselves in an ordered manner such that there is net dipole moment in the crystals or (ii) they may align themselves in such a manner that the dipole moments cancel each other. (iii) It is also possible that there are no polar species (dipoles) in the crystal, but only ions are present. Crystals where the situation (i) is found, exhibit piezoelectricity. When such a crystal is deformed by mechanical stress, electricity is produced due to the asymmetric displacement of ions or conversely if an electric field is applied to the crystal, there will be atomic displacements causing mechanical strain. Thus, a piezoelectric crystal acts as a mechanical-electrical transducer. Some of the polar crystals, when heated, produce a small electric current or pyroelectricity due to the asymmetric variation in interatomic distances.

In some of the piezoelectric crystals, the dipoles are permanently lined up (spontaneously polarized) even in the absence of an electric field and the direction of polarization can be changed by applying an electric field. This phenomenon is called ferroelectricity by analogy with ferromagnetism. Barium titanate ($BaTO_3$), sodium-potassium tartrate (Rochelle salt), sodium nitrite ($NaNO_2$) and potassium dihydrogen phosphate (KH_2PO_4) are typical ferroelectric solids.

The occurrence of ferroelectricity undoubtedly depends on the structure and only certain geometric arrangements in the crystal can exhibit this phenomenon. If the dipoles in alternate polyhedra point up and down, there will be no net dipole moment and the crystal is said to be antiferroelectric. Lead zirconate ($PbZrO_3$) is a typical antiferroelectric. Both ferroelectric and antiferroelectric crystals transform at some temperature to the paraelectric phase where the dipoles are randomly oriented.

NON-CRYSTALLINE SOLID STATE

The liquid phase is the state of substances at temperature above their melting points. This includes all amorphous substances (silicate glass, rosin, etc.) Since silicate glass has no crystalline lattice, all amorphous solids are termed as glassy or glasses. Therefore, the two phase states of solids are crystalline and amorphous (glass). The glass transition is the passage of a molten liquid into the solid state without changing the phase, *i.e.*, with the retention of short range order. When the liquid is cooled at a fast rate, its coefficient of viscosity increases and its energy of thermal motion decreases. The molecules remain in the random disorder characteristic of a liquid. Since it involves the breaking and reforming of strong covalent bonds, there is only a short-range order and a long-range disorder ultimately at a certain temperature at which B the coefficient of viscosity approaches 10^{13} poises which correspond to viscosity coefficient of solids. So the super-cooled liquid solidifies and the substance passes into amorphous (glassy) state. There are two types of substances that can form glasses.

Glass

Substances, whose liquids consist of long molecules, that become very much entangled, form glasses. Organic polymers and polymeric form of sulphur, selenium, belong to this class.

Open Structure Crystals

These substances, which have directed covalent bonds and low coordination number, crystallizes in an open structure. SiO_2, H_2O, oxides of B, Si, Ge, P and As belong to this class. H_2O glasses are formed on extremely rapid cooling or upon condensation at 100°C. Alcohols, glycols and some other hydroxy substance also forms glasses. In the case of SiO_2, each silicon atom is surrounded tetrahedrally by four oxygen atoms, each of which is bonded to two silicon atoms. The bond length of Si-O is 1.62 A and the ratio of the distance between comers of tetrahedron to the distance from a comer to its centre is 1.64. In the melt Si–O bonds are continually forming and breaking.

When the melt is rapidly cooled the particular bond pattern present at that instant is frozen. The resulting glass is still a giant molecule but a rather disordered one. The result is a solid with the rigidity of a crystal but no regular order. In glassy state the tetrahedron order is distorted. The positional disorder existing within each distorted tetrahedron prevents the prediction and assignment of fixed position to the atom.

NON-CRYSTALLINE POLYMERS

Polymers undergo continuous transition from an ideal order to an absolutely disordered state. Very many polymers do not exhibit crystalline state and the crystalline order is also not found in the melt and the solutions of polymers because the thermal motion destroys the crystalline lattice. In the case of rapidly cooling the polymer solution, either a completely amorphous polymer or a body where crystalline and amonaphous region coexist, can be obtained. All the above mentioned cases form a group of non-crystalline bodies eronously called amorphous polymers.

Kargin suggested a definite degree of ordering within the limits of a non-crystalline state. He proposed a bundle model containing a set of macromolecules which are packed almost parallel and have large longitudinal dimensions. Let W(R) be the relative probability W of finding neighbouring atoms at a distance R from a certain fixed atom.

For a crystalline substance this function shows constant distances at which the neighbouring atoms are arranged, exhibiting a long range order in the body. For non-crystalline polymers, it completely changes. The first high maximum conforms to the probability of an ordered arrangement of the first particle layer relative to the fixed particle showing a short range order. Such an arrangement results in ordered clusters on associates. Since the height of subsequent maxim is considerably lower, W(R) –1 and hence, the long range order is absent.

In the case of non-crystalline polymers the X-ray patterns give d_1 as 4.5-5Å and d_2 as 10Å. The presence of these maxima show that either macromolecules or segments of macromolecules are packed almost parallel to one another. Therefore, the main feature of non-crystalline polymers is a short-range order up to 10Å and absence of a long-range order. The employment of electron microscope has shown that in crystalline polymers folding chains form the crystalline lamella, but in amorphous polymers parallel arrangement of chains occur up to 30-100Å, showing a short-range order. There is great deviation from a crystallographic order in the domain.

It is assumed that these domains are connected with another by tie chains. Such domain exhibit transition from an amorphous to a crystalline structure. The non-crystalline polymers display a truely amorphous region which are formed by chain ends. Yet has built a model for amorphous polymers showing domains, the chains, and truely amorphous regions. Kahanon postulated the formation of super domain structure in amorphous polymers.

Plastics

Polymers according to their mechanical properties have been classified as elastomers and plastomers. Elastomers are rubbery elastic materials. Plastomers or plastics are macromolecular materials which flow on deformation. The deformation is permanent in plastics but in elastomers the body returns to its initial state instantaneously upon the removal of stress. Normally, in elastomers the chains are very flexible at the room temperature but they are stiff in the case of plastics. This depends on several factors namely, the degree of polymerization, cohesive forces acting between the chains, flexibility of the chains, their branching and cross linking and the degree of crystallinity. The effect of crystalline structure is discussed below.

Polymer Crystals

Polymers are products of high molecular mass and their decomposition temperatures are far below their boiling points. Hence, they exist only in liquid or solid state. Although most of the polymers are amorphous, but the crystalline polymers known as isotactic polymers have high melting point and produce fibres of high tensile strength and excellent mechanical properties. A polymer has two types of structural elements viz. monomer units and chains. The long range order is a prerequisite for the crystallization of polymer, not only in the arrangement of units but of chains as well as in three dimensions. This can happen in three ways:

1. Long-range order of arrangement of both chains and monomeric units in three dimensions.
2. Long-range order of arrangement of chains, but the monomeric units are arranged disorderly.
3. Long-range order of the arrangement of units but short range order of chain arrangement.

The high degree of ordering can be achieved in polymers in two ways, *i.e.*, by crystallization or by chain orientation without orientation of monomeric units. These correspond to crystalline and amorphous (glassy) state. It is possible to have a perfect short-range order in chains but the long-range order in arrangement of chains and monomeric units is not very perfect. Therefore, the examples of well ordered liquids and defective crystals are well known.

Crystallization of Polymers

Many polymers are incapable of crystallization under any condition. Their crystallizability is determined by their chemical constitution, *i.e.*, their packing, energy of intermolecular reaction, chain regularity, flexibility, etc.,

Regularity of Polymer Chain

A long-range order must exist in the chain in three dimensions. Sometimes irregularity in the chain may not obstruct crystallization but random and atactic copolymers cannot be crystallized under any circumstances. Natta by using certain catalysts succeeded in synthesizing several highly crystalline vinyl polymers, *e.g.*, polypropylene, polybutylene and polystyrene.

A mixture of the polymers of various degree of orientation was formed and crystalline fraction was separated from amorphous fraction by means of solvents as crystalline polymers are less soluble than amorphous polymers.

Second Order Transition Temperature

A first order transition is one at which a definite change in volume occurs. A second order transition is one at which a change occurs only in the rate of change of volume with temperature. Below T_g, polymers are rigid, glass like material and above T_g they are soft elastometric materials. The second order transitions are caused by thermal agitation releasing portions of the resin chain to rotate around its bonds. These second order transitions can be measured conveniently by deflection temperature DT tests.

Flexibility of Polymer Chain

Polymers with sufficiently flexible chain can crystallize, but liquid chain polymers crystallize with difficulty because the thermal motion of the momeric, units disturb the established order and the crystal breaks. To crystallize a polymer, it is cooled to a temperature where the motion of its units does not hinder their orientation but excessive cooling may also hinder the rearrangement of the units. Crystallization of each polymer is possible only within a definite temperature interval, *i.e.*, between the glass temperature and flow temperature, *e.g.*, the glass temperature of isotactic polystyrene is about 100°C. Hence, it will only crystallize between 100°-220°C which is its melt temperature.

Packing of Molecules

Polymers obey the principle of close packing and macromolecules must be packed as closely as possible in the crystalline lattice. There are several possibilities for the formation of close packings of polymer chains. The first crystalline structure is built by close packing of spheres which is exhibited by globular proteins. The formation of such crystalline structures are possible because all the spheres are equal in size, *e.g.*, edestion catalase, iodobenzoyl glycogen.

The second possibility of close packing is the packing of helical macromolecules. The convexities of one helix fit into the concavity of other, *e.g.*, amylose, poly (hexamethylure adipamide), polypeptides. The third is the packing of long extended chains, *e.g.*, polyamide, polyurethanes, etc. The essential requirement is that the side chain substituents must not hinder regular arrangement of adjacent chains.

Close packing of chain molecules can only be accomplished with flexible chains capable of shifting in parts. Introduction of polar groups gives rise to two opposite effects. On one hand, intramolecular attraction increases, favouring close packing, while on the other hand, the thermodynamic flexibility (potential energy difference between two positions of the chain) of the chain decreases.

The chain becomes more rigid. The rate of close packing depends on these two factors. However, with polar groups like –OH or –CN, the attraction effect prevails inspite the fact that the chains are rigid. Since rigid chains closely pack in the extended state, such polymers are highly oriented, *e.g.*, poly (vinylalchol), polyacrylo butyrate. The chains of isotactic polypropylene are highly flexible and form crystalline lattice, but rigid cellulose chains cannot form such lattices under any condition.

Polymer Crystal Cells

The crystalline cell of polymers consists of parts of chains located at definite distances a, b, and c and angles α, β and γ. They do not differ at all from cells formed by low molecular weight compounds and obey all the rules of symmetry.

Polyethylene crystals form orthorhombic unit cells with a = 7.40Å, b = 4.93Å and c = 2.53Å and density equal to 1.0 g/cm^3. All the carbon atoms in the polyethylene cell are arranged on the same plane forming a zigzag plane. In the polyethylene unit cell, chains are arranged along four edges and in the centre of the cell.

In the case of chains having bulky side substituents helical crystals are formed, *e.g.*, polypeptides, polypropylene divinyl and olefin polymers. One helical loop may contain a different number of monomeric units depending on the nature of a polymer. Natta synthesized 1,2 polybutadiene and similar results were experienced with isotactic polystryrine.

Polymers like sulphur or phosphorus are characterized by polymorphism. Polypropylene may form crystals belonging to monoclinic, hexagonal and triclinic types of symmetry. They are typical first order phase transitions. These transitions between various polymorphous formations take place either upon a change in temperature or under the action of load which brings a change in the parameters of a crystalline cell. As a matter of fact most of the polymers can be crystallized under stretched conditions. The morphology of a crystalline polymer is determined by the mutual arrangement of the unit cells within the limits of the crystalline state of a body.

Single Crystals

Single or ideal crystals are formed by the parallel transfer of unit cells along edges at distances equal to periods. Fischer first of all ascertained the possibility of high molecular compounds to form monocrystals. Polymer single crystals like other crystals have been found to contain certain defects originating in the packing of chains.

Lamellar Crystals

Polymer crystals are grown by crystallizing from a one per cent solution of polymer under gradual cooling or isothermal aging under the temperature of dilution. The dimensions, shape and regularity of structure of a single crystal depend on the chemical structure of a chain and the physical constraints like temperature, concentration of solution, the cooling rate, the nature of the solvent, etc. The polymer single crystals are monolayer flat plane Lamella which are often rhomboid of 100Å thick and the size of the plate up 1 mp. The axes a and & of a crystalline cell correspond to long and short diagonal of the rhomboid, while axis c along which the polymer chains are directed is perpendicular to the crystal plane.

Since the length of the polymer is very big while thickness of the crystal is of the order of about 100Å, the long chain can only fit in the crystal by turning on its surface by 180°. This is known as chain fold conformation. To have an idea in a crystal of 120Å thickness, the fold shall contain 100 carbon atoms. A molecule having a weight of 105 shall fold about 70 times.

The thickness of a lamella, the dimensions of crystallite and degree of defects of the boundary layers of a crystal are affected by the crystallization conditions. Crystals with folded chain conformation can be obtained when super cooling is in the range 15° – 20°C.

Crystallization under high hydrostatic pressure 5-10 × 10^3 atm give extended chain crystals. Polyethylene annealed under a pressure of 700 atm give extended chain crystals. When the melt is cooled under great deformation crystals of different morphological forms can be obtained. Polyhedrons known as hedrites and axialites or avoids are obtained by raising the concentration of the solutions.

Fibrillar Crystals

Fibrillar crystals are obtained by the high rate of evaporation of a solvent from a moderately concentrated solution or at cooling of melt. There are two views about their formation. One section believes that aggregation of rolled lamella is responsible for them while other hold that lamella degenerate in the formation of fibrillar crystals.

In stereospecific polymerization, fibrillar crystals are directly found. The formation is determined by the ratio of the rates of growth of a chain and its crystallization. By changing this ratio crystal either with extended or folded chains are obtained.

Globular Crystals

In globular crystals, macromolecules in coiled conformation form the lattice points. These are found in biopolymers, as a very high degree of uniformity of macromolecules is a pre-requisite for it, *e.g.*, tobacco mosaic virus synthetic polymers do not form them.

Spherulites

Spherulites are obtained when polymers crystallize from melts or concentrated solutions. They are made of crystallites which grow radially from one common centre. Macromolecular chains make at least an angle of 60° with the radius, *i.e.*, they are arranged tangentially to the spherulite radius Spherulites possess anisotropic properties. Their indices of refraction of light are different in radial or tangential directions. The orientation of the crystallographic axes continuously changes along the angular coordinates, A spherulite is called positive when the refractive index in the radial direction is greater than in the tangential one; otherwise it will be a negative spherulite.

In radial Spherulites one of the axis of the crystalline lattice retains a constant direction along all the radial directions. Ring type Spherulites are built of lamellar crystals whose orientation continuously changes along the spherulite radius. Depending on the degree of cooling, the same polymer may form different types of spherulitis. With low degrees of super cooling ring type Spherulites are obtained while radial Spherulites are obtained by high ones.

The formation of Spherulites passes through several stages. First the crystal nuclei are formed which are scattered throughout the volume of the sample and then independent growth of discrete fibrillar or lamellar crystalline structure grow at the same time in all directions. Then encountering one another they grow into regions filling the entire volumes of the body. Then their boundaries are distorted and they assume the form of a polyhedron.

The structural units in spherulite are bundles of macromolecules made out of chains in an extended conformation. This is possible when several macromolecules try to crystallize simultaneously from both the ends. Such interstructural bonds combine to form separate crystallites inside fibrillar or lamellar crystals.

Oriented State of Polymers

Polymers because of great anisotropy of the dimensions of macromolecules exist in oriented state. In the extreme case all the macromolecules will be oriented parallel to each other, but in reality this can happen on the average as the flexible chains do retain some fold conformations. In the oriented state the difference in amorphous and crystalline polymer vanishes.

Peterlin suggested that in oriented polymers microfibrils of diameter 100-200Å clearly divide crystalline blocks of chains with fold conformation. These chains are divided by defective layers and a large number of intra- and smaller number of interfibrillar chains. The microfibrils continue into larger fibrils to create the macrostructure of oriented bodies of the size 1000Å. These polymers are important for producing the fibres and films. In the first case an uniaxial orientation takes place while in the second a planar orientation takes place.

Liquid Crystals

Certain substances that contain rod-shaped molecules can exist between the crystalline and liquid state. The forces between these molecules are greatest when they-lie parallel and the crystals of such substances consist of parallel arrays of molecules stacked side by side when the crystal is heated, a point is reached at which the thermal energy is sufficient to break down the crystal lattice but not the parallel arrangement of the molecules. This results into a liquid crystal when the molecules are free to move provided they all remain parallel. At higher temperature the liquid crystal melts to a liquid. Such crystals are called *nematic*.

There are other liquid crystals where the molecules are not only parallel but their ends are aligned. Such crystals are known as *sematic.* In this case the crystals contain layers where the parallel molecules are free to move about in its layer but not in other layers. These crystals form drops in which the surface forms sharp steps with flat risers in between.

The intermolecular attraction in liquid crystals do not act equally in all directions with the result that all the degrees of freedom are not liberated. In a nematic liquid crystal the molecule can rotate only about its long axis and two rotational degrees of freedom are lost. In sematic liquid crystal the molecules are free to move only in the two directions and some translation degree of freedom is lost.

ELASTOMERS

Elastomers are non-crystalline high polymers that have a three dimensional space network structure.

They are built up like a fishing net, the siring of each mesh is replaced by a flexible chains of atoms with only weak forces between separate chains. These chains are prevented from melting by crystalline or by covalent interbonded portions. The distance between the head and tail of the chain is greatest when it is stretched out straight and less when it coils itself. It is therefore, elastic at the room temperature. Unlike steel, stretched elastomer is found to shrink on heating. It is another example of ordered and disordered arrangement.

UNIT CELL OR LATTICE

A unit cell is a convenient block, carved out from a crystal of infinite extension, bound by three pairs of parallel planes. The movement of this block along the three axes reproduces the whole crystal.

Liquids on cooling give out energy, resulting in the decrease of translational energy of molecules, so much so, that they come under the influence of the field of secondary valency forces. The energy at this stage is only sufficient for rotational and vibrational motions of molecules, atoms, or ions, about a position of maximum potential energy, resulting in a solid.

Liquid Crystal Phase Transition

Lennard-Jones *et al.,* developed a model to explain the phase transition between the liquid state and solid crystalline state. They assumed that solids are constituted of two interpenetrating lattices represented by a-sites (+) and β-sites (0). In a perfectly ordered state, all the atoms occupy α-sites and holes occupy β-sites. If X is the fraction of atoms on α-sites, the long orders will be defined as 2X-1. With increase in temperature some atoms will migrate to β-sites, which are at a higher potential energy due to repulsive forces that are predominant at distances less than the equilibrium distance between α-sites.

A few such displacements will still retain long range order. At sufficiently high temperatures, the atoms will be randomly distributed between α-sites and β-sites and the long range order will completely disappear. The lattice retains a perfect order with all the atoms occupying α-sites even at high pressure over a wide range of temperature. The lattice can expand without loosing its order provided the kinetic energy of the molecules and the repulsive force between them decreases monotonically with increasing volume at constant temperature.

This increase in volume will create more disorder because of the movement of molecules to β-sites. The initial increase in disorder will be responsible for the increase in pressure due to the increase in repulsive force but when the volume becomes large enough, it would be followed by decrease in pressure as it would permit easy movement of molecules to β-sites. Thus, this model suggest a phase transition at constant pressure, leading to a critical temperature like gases, above which a liquid-solid transition does not occur. This phenomenon has not yet been observed even at highest pressures attainable.

Crystalline Solid

The molecules, under the influence of cohesive force, will try to orient themselves in an ordered manner in space, provided the rate of cooling is so adjusted that the molecules get enough time to orient and the density of the liquid is not too high. Thus, a crystalline structure will result. Since there is only a single spatial arrangement of minimum potential energy for all the molecules in a compound, therefore, in the transition from liquid to solid state, the same spatial arrangement is repeated, *i.e.*, we obtain a crystalline lattice of the molecules.

Lattices and Structures

A regular array of points in space is called a lattice.

If identical atoms or molecules occupy fixed position in a lattice, a crystal structure is obtained.

Closed Packed Spheres

The spheres of the same size can be closely packed in one plane. They are so arranged that their centres occupy the comers of an equilateral triangle and each sphere is surrounded by six other spheres forming a regular hexagon, *i.e.*, their coordination number is six. The various depression around each sphere are known as interstitial sites, void spaces or holes.

Let us place the second closed packed structure of spheres over the first layer such that small open circles show the position of the centres of the spheres of lower layer and the shaded small circles shown the centres of the spheres of upper layer.

The centres of the upper layer will also occupy the comers of an equilateral triangle. Now the third layer can be placed either at the centre of the first layer or at the depressions of the first layer. The second and the third layer will also be closely packed.

Hexagonal Closest Packing (HCP)

Place the third layer at the centres of first layer and the holes of the second layer. We will obtain a structure which is known as hexagonal closest packed structure.

Cubic Closest Packing (CCP)

Place the third layer at the vacant sites of the first layer. Then we will obtain a structure. This is known as cubic closest packing. Each interstitial site is surrounded by four spheres.

Interstitial Sites

Supposed atoms form a close packed structure and B occupies the vacant sites. It will form a triangular coordination. In the packed layers if one atom A will be over other three atoms and the centres of the four spheres will be on a regular tetrahedron, then the space at the centre of this would be known as tetrahedral site. Thus, the holes above each sphere of the lower layer will give a tetrahedral site. In the case of unfilled holes of the first layer, the site is surrounded by six spheres whose centres would be at the comers of a regular octahedron. This site is known as octahedral site.

In case the centre of any regular tetrahedron falls on equitorial plane formed by the centres of four closely packed spheres, it will give one octahedral site for every sphere in the lattice. All the above conclusions will hold good for hexagonal closed packed structures also.

Let the lattice points A, B, C, D, E, F & G be at the corners of a cube. The only planes of interest are those containing lattice points. These can be divided into groups, according to the intersection which the planes make with the rectangular axes X, Y and Z.

The intercepts along the axes are spaced as a, b and c respectively. These are known as Weiss indices. A more convenient method is to describe the intercept along the axes by a/h, b/k and c/*l* respectively. The integers h, k and *l* are known as Miller indices of the planes. In case of fractional intercept, they are multiplied by the smallest common factor to convert all the three into integers.

In a cube, since a = b = c = 1, the plane containing the points A, F, G, E cuts ox line at A and the other axes at infinity, it is called a (1,00) plane; similarly, the plane containing the points A, B and C is called (1, 1, 1) plane.

Angles between the planes. The line OA of length d_{hkl} is the line from the origin perpendicular to the nearest hk*l* plane, a is the angle between OA and the axis X, β is the angle between OA and the axis y, γ is the angle between OA and Z axis. Thus, having known the lengths a, b and c and the angles α, β and γ, a lattice can be fully described.

TYPES OF LATTICE

It can be understood from our knowledge of closest packing that the simplest lattices will be cubic and hexagonal.

Cubic Lattices

There are three types of cubic lattice namely simple cubic lattice, face centre cubic lattice and body centred cubic lattice.

Simple Cubic Lattice

With the exception of the halides of cesium, all alkali halides gives crystals in simple cubic lattice. In this lattice all the nearest neighbours of any given ion are six ions of opposite charge, each at a distance of 1/2 from the central ion. In case these ions are spheres we find those at the centres of the faces are half outside and those at the comers are only one-eighth inside. Thus, the unit cell will contain $1 + \left(12 \times \frac{1}{4}\right) = 4$ alkali ions and $\left(6 \times \frac{1}{2}\right) + \left(8 \times \frac{1}{8}\right)$ halide ions.

Face-centred Cubic Lattice

All the rare gases in solid form arrange themselves in face- centred cubic lattices. Cu, Ag, Au, Ca, Sn, and Al and strontium also belong to this class. The unit cell contains four atoms, each of which is surrounded by 12 neighbours at an equal distance of $\frac{a}{\sqrt{2}}$ and the radius of the atom will be $\frac{a}{2\sqrt{2}}$

Body-centred Cubic Lattice

Li, Na, K, Rb, Cs, and Ba give a crystallization body centred cubic lattice. The central ion is surrounded by eight equidistant neighbours at a separation of $\sqrt{3a/2}$. The radius r will be $\sqrt{3a/4}$.

METALS

Non Metals

Non-metallic elements form maximum number of natural and synthetic compounds. We shall try to study the elements in first few series in different groups with the help of ball models.

Allotropes

The existence of an element in more than one form in the same physical state is known as allotropy and the elements are known as allotropes, *e.g.*, o- and p-hydrogen, He-3 and He-4, etc.

Hydrogen

Hydrogen molecule exists in two allotropic forms known as ortho and para hydrogen. Arrange the balls. Let us study the effect of rotation of the balls on their axes. There are two ways in which the balls can rotate on their axes. A pair of balls will either rotate in the same direction or in opposite direction case these balls are assumed as the positive charges (protons) of hydrogen atom. Will represent ortho hydrogen and will represent para hydrogen (the electron going around the proton.

Table 2

Shape	$a_0 Å$	$C_0 Å$
Tetragonal	4.42	3.75
	4.50	3.68
	5.40	5.08
HCP	3.78	6.16
	3.75	6.495

Hydrogen liquefies at 252.77°C and solidifies at 259.2°C at 55 mm pressure. The crystalline structure of hydrogen depends on the pressure. Under extreme pressure, metallic hydrogen is obtained. Hydrogen gives isotropic crystals belonging to cubic system shapes of either octahedron or hexoctahedron with needle like structure. Table 4.11 gives data about hydrogen crystals by various workers.

Helium

Helium is an inert gas and has two isotopes in the liquid form viz. He-3 and He-4.

Liquid State : Helium 4 on being cooled below 2.172°K gives a super cooled liquid. The transition to super fluid state is not a first order transition. The phase diagram shows a sharp peak resembling δ. Hence, it is known as δ point.

Normal helium 4 liquid is known as helium I and the super fluid as II. He I & He II show a ccp structure at 4.2°K to 4.50°K with 12 atoms at 3.2 A apart. Helium-3, displays three super fluid phases. The phase diagram for liquid and solid mixtures of He-3 and He-4.

Carbon

Carbon is the constituent of numerous naturally occurring substances and also forms the very basis of all organic compounds.

Allotropes of Carbon : Carbon exists in several allotropic modifications, crystalline (diamond Fullerene and graphite) as well as amorphous.

Graphite : Graphite exists as black slippery crystals. Bemal describes the structure of graphite as hexagonal. Infinite layers of atoms of carbon are arranged in the form of hexagons lying in one plane. They are stacked as ABAB so that the atoms of every third layer align with each other.

The minimum distance between each layer is 2.44 A. Lipson has shown that graphite also gives rhombohedral crystals with ABCABC stacking arrangement so that every fourth atom in the layer align with each other. The distance between each carbon atom is 1.415Å and the distance between each layer is 0.3354 nm.

Diamond

Carbon atoms in diamond crystals is covalently attached to four carbon atoms placed at comers of regular tetrahedron. C-C distance is 0.155 nm. Diamond gives a face-centred cubic structure. The wrinkled sheets lie in the octahedral face plane of a crystal stacked in ABCABC sequence. A well crystallized graphite can be converted to diamond structure by applying a pressure of 130 K bar. It will give diamond crystal in 0 hexagonal form with ABAB sequence and will have four atoms per unit cell.

Fullerenes or Bucky Balls

Fullerene are the third crystalline modification of carbon. They are a special class of carbon clusters and formed by laser vapourisation of graphite in helium atmosphere. These clusters differ from one another in the number of carbon atoms and in their configuration. Since these structures are similar to 'geodestic dome' invented by Buckminister Fuller, they were named as buckyball or fullerene. These structures are made up of combination of pentagons and hexagons. They differ in the manner they are combined.

C_{14} and C_{16} are the earlier stable clusters, C_{14} is made up of two fused pentagons giving rise to a octagon which combines with a hexagon, giving octagon type of structure. C_{16} is made up of two pentagons combined with one hexagon, giving a perfect ball structure.

C_{60} finds a unique position. It is made up of 12 pentagons and 20 hexagons, giving a perfect ball like structure. The relative position of carbon atoms in the ball, each and every carbon atom is exactly identical to that of other 59 carbon atoms. This arrangement of carbon atoms imparts highest degree of symmetry in the third dimension resulting in highest stability. Euller has also proved the above mathematically.

Formation

The important question is that how does 60 carbon atoms come together? Evanston of Northern University has suggested the following mechanism:

The carbon atoms first form paired hexagonal rings around which a long chain of carbon atoms extends. These paired structures close up in a process of cyclization and form fullerene fragments. This fragment serves as the seed structure along which dangling carbon chain link themselves into pentagons and hexagons. Thus, the chain spirals around the seed structure, forms a spherodial fullerene structure.

C_{70} fullerene is made up of 12 pentagons and 25 hexagons. They contained a rugby shaped ball. The structure of higher fullerenes. It is observed that more the structure is away from C_{60}, the more it looses its stability, therefore, C_{70} is less stable than C_{60}.

Properties

C_{60} fullerene is the purest form of carbon. Both diamond and graphite on being cut give surfaces that are instantly covered with hydrogen which ties up the dangling surface bonds:

Chemical properties of fullerenes can be discussed in two divisions, namely, their oxidation/reduction properties and its functionalization chemistry. The electron transfer to fullerene is reversible to such an extent that it may be termed as an 'electron sponge'. To explain, C_{60} has a tremendous electron affinity, so that, it can accept upto six electrons, both chemically and electrochemically.

Resulting anions, C^{-}_{60}, etc. are remarkably stable and can form radical ion pairs with the donor (D) cation, for example, $[C_{60}]$ [D]. Such ion pairs may exist as stable charge transfer salts. The extra electrons can be easily drawn out from the anion as easily as a sponage is wrung out. C_{60} can be oxidized by super acids Uke 'magic acid', (FSC_3 H: $SbCl_5$) or fluorosulphuric acid (FSO_3 H) alone. But the resulting cations are highly unstable because of the inherent electronegativity of it.

About 12,500 resonating forms are possible for buckyball; still it is less aromatic than benzene. Considerable studies have been done regarding the preparation of derivatives of C_{60} via numerous routes. These derivatives possess unique physical and chemical properties, which differ substantially from parent fullerenes.

For example, buckyball has been hydroxylated to include upto twenty hydroxy (OH) groups in its structure. This compound, 'Fullerol', is water soluble. Similarly, extensively fluorinated C_{60} F_{60} schemes have a unique lubricating property, so that, it is termed as 'Molecular Teflon'. Various additive compounds, expoxides, organometallic complexes, etc. of buckyball are well known. Some of these derivatives exist in a few isomeric forms which differ from each other in physical and chemical properties.

Fullerenes have made their way into the world of chemistry through a different route, which relates their properties with polymers, polymerized by the UV irradiation process in absence of oxygen or by plasma discharge, into $(C_{60})n$ where the value of n is around 20. But more interesting is to insert buckyball cages into long polymeric chains of hydrocarbons, like poly (1,3-butadiene) to make a 'pearl necklace' structure. In another way, C_{60} cages are made to dangle like pendants from a polymeric chain, resulting in a 'charm bracelet' polymer.

Starting from reactive hydoxyl (OH) groups of extensively hydroxylated 'fullerol', same or different polymeric chains can be attached to the C_{60} cage. These are called 'star burst' polymers or 'dendrimers'. A growing polymer chain can often be 'end capped' with terminal C_{60} cages. 'Grafting' of C_{60} onto polyethylene surface has also been reported. These polymer analogues of fullerenes have brought newer promises for the technological world and prospects for better utilization of fullerenes.

Apart from these chemical modifications, scientists were wrestling with the question of how to utilize buckyball's empty interior. Kroto, Smalley, and Coworkers are pioneers in this branch also. They have succeeded in imprisoning potassium atom and caesium atom into the C_{60} cage using laser vapourization technique. Scientists have also been able to encapsulate different metal atoms of varying sizes and reactivity namely calcium, barium, lanthanum, yttrium, uranium and so on, along with some molecules also.

However, all these encapsulations have been performed without significantly disturbing the structure and property of the fullerene molecules. Properties of these entrapped molecules while 'residing' inside the buckyball cage opening the door to a new 'nanoworid'. ('Nano' is a word used to describe matter at the scale of 10^{-9} m or one thousand millionth of a meter.). Whether a fullerene cavity can serve as the smallest vessel in the world to trap a reactive atom in its atomic state is also a very interesting subject for scientists.

In recent days scientists are getting interested in the novel electrical and electronic properties of fullerenes. In different compounds, fullerene changes its property from being an insulator to turn into a superconductor! To explain, buckyball itself is an insulator in its pure form (band gap

= 2.2 ev). But, it becomes a semiconductor when suitably 'doped'. This doping may involve putting a dopant atom or molecule inside the cage or some other appropriate route forming 'buckide' salts.

'Fulleride' or 'buckide' salts conduct electricity; conductivity increase with increased metal doping and reaches a maxima at a C_{60} metal ratio of 1:3. With further increasing metal doping the buckide salts start to lose conductivity, $M_6 C_{60}$ (M is a metal atom) being an insulator. What is more important is that $M_3 C_{60}$ is a high temperature superconductor, with a Tc of 18K to 28K. Mixed doping with Cs and Rb results in the formation of a superconductor $Cs_2 RbC_{60}$, with a Tc of 33K.

However, all these buckide salts are three dimensional superconductors as compared to previously known two dimensional cuprate superconductors. Discovery of superconductivity in buckides has given a new impetus to fullerene research. Apart from being 'doped', buckyball itself can dope organic conductors/semiconductors. For example, organic polymers like polyaniline, polyacetylene, etc., can be doped with buckyball to induce semiconductivity in them. These polymers, when otherwise doped, behave as 2D semiconductor. In India, DR C.N.R. Rao of the Jawaharalal Nehru Centre for Advanced Research, Bangalore and his group is also working with superconductivity in fullerides.

Four different types of fullerene compounds have so far been discussed, namely the charge transfer (CT) salts, chemically modified derivatives, interstitially doped metallofullerenes, and the buckide salts. CT salts account for the First 3D organic conductors, several high, Te superconductors, and organic ferromagnets with a specified Curie temperature. Buckyball's ability to trap different atoms and molecules has successful application in medical and biomedical fields.

For example, radioisotope encapsulated C_{60} has successful application in cancer therapy. HIV protease, an enzyme essential for growth of AIDS virus HIV, has a diameter almost equal to that of the C_{60} cage; therefore, the capability of C_{60} to encapsulate HIV protease may have immense importance in AIDS therapy. Similarly, the interstitial cages of C_{60} and other fullerenes are being utilized for various purposes such as solvent preservation, controlled drug release, etc. Important biomedical application of water soluble fullerols has also been suggested.

Important applications of catalytic amounts of buckyball have been found. It can improve antiwear, antiseize, antifriction properties of lubricating oils. In petrochemical refining industries, efficient production of many hydrocarbon products are often hindered by undesirable polymerization of reactive olefins. Fullerene, in ppm (parts per million) level, can inhibit such thermal polymerizations and thereby improve the refining efficiency. Dissolved Fullerene inhibits coke forming on a catalyst and may therefore, be used as an anticoking agent. Preparation of glassy thin Films containing C_{60} could offer the possibility of producing filters against optical damage.

The boundary between theory on a catalyst and practise is becoming thinner as the world of fullerenes is being explored. However, all these promises may die within the four walls of laboratory, until and unless the buckyball can be synthesised in plenty.

In view of Dr Kroto, C_{60} should exist in abundance in interstellar space, arising from chemical reactions in the outer atmospheres of carbon rich red giant stars. The prediction has been evaluated by spectroscopic observations of astrophysicist Webstar of Edinburgh Royal Observatory. Besides, trace amounts of C_{60} have been found in coal deposits and in some geological stone samples. In 1966, Dareid F. Jones, published in a paper 'New Scientist' suggesting that the fullerene C_{60} is made up of curbed graphite sheet. But Leonard Euller has proved that it is not possible. Sumito ijima has suggested that curbed graphite sheet will give a cylindrical tube.

These are hollow, three dimensional layered structures called nanotubes. These are of the order of ten to a thousand or more nanometers in length, a few nanometers in diameter and posses several tens of such layer. One side of it is closed with fullerene type of hemispheres. They are stiffer than any other known materials.

They serve as conducting nanowires when stuffed with molten conductor. It is expected to use them for chemical and molecular reactions inside these test tubes. In the synthesis of nanotubes, hollow polyhedral particles of diameters 5.50 mm are produced. Motallow fullerene like lanthanum carbide (LaC) are produced. The encapsulation into nanoparticles or nanotubes may prove to be a useful method for protecting sensitive materials. Fullerene and nanoparticles form 'Giant Fullerene'. These have a diameter of 1.2 mm. Thus, these have opened a new and very wide field of research.

Amorphous Carbon

Amorphous carbon exists in many forms like coal, lignite, gelsonite, coke, gas carbon, charcoal (animal, wood, sugar), lamp black and soot.

All these are characterized by imperfect graphite structure. The parallel layer planes are not overlaid to their perpendicular axis, but the angular displacement of one layer with respect to the other is random and the layers overlap one another irregularly. This is known as turbostratic structure. The distance of separation is of the order of 0.35 to 0.365 nm and C-C distance is 0.142 nmboth in graphite and amorphous carbon.

Nitrogen

Nitrogen is an inert gas and does not form polymers in the elemental form. Ammonium cyanide NH_4CN at –125°C forms isochains.

Nitrogen liquefies at –195.8°C and solidifies at –209°C. In solid form nitrogen exists in two allotropic forms namely a-nitrogen and α-nitrogen.

α-nitrogen gives F.C.C. crystals at 4.2°K with a = 5.644 A and Z= 4.

β-nitrogen give hexagonal crystals at –209.9°C with a = 4.039 A and C = 6.67 A. N-N distance is 1.067 A.

Oxygen

Oxygen is the most reactive element and exists in two allotropic forms namely O_2 and O_3 (ozone).

Oxygen liquefies at –182.5°C. Dewar obtaineda hard pale blue solid by cooling liquid oxygen to –227°C at 0.9 mm pressure. In the solid form oxygen gives three allotropic forms namely α–, β–, γ oxygen.

α-oxygen-249.5°C

β-oxygen-230.5°C

γ-oxygen.

Oxygen crystallizes in hexagonal shape when crystallized in boiling hydrogen.

Ozone boils at –111.9°C and solidifies at –192.7°C, but the structure of its crystals is not well established. A modification of ozone consists of oxygen isochains.

Halogens

Of the four elements offhalogen family found in nature, fluorine and chlorine are gases, bromine is a liquid and iodine is solid. Halogens in the solid form given othorhombic crystals and all atoms exist in tetramolecular units.

Fluorine exists as a yellowish orange liquid at –188°C, yellowish solid at 220°C and white solid at –228°C. It given mono clinic crystals at –233°C.

The chlorine and other halogen atoms form layers parallel to 0/0 axis. In I_2 the molecules are grouped in layers parallel to ac plane. There are 16 chlorine atoms in a tetragonal unit. Each chlorine atom has two neighbours at a distance of 3.34Å. The distance between Cl-Cl layers is 3.69Å. The atomic separation between bromine molecules is 3.30Å. Rapid evaporation of rhombic bromine crystals give monoclinic crystals.

Silicon

Silicon is inert at room temperature. At the atmospheric pressure, its crystal lattice is the same as that of diamond, *i.e.*, two interpenetrated face-centred cubes. At high pressures but less than 15000 aim., the FCC structure is converted to BCC structure. Vapours of silicon at 500°C on cooling give amorphous silicon. The crystals grown from vapours give octahedral structure. Silicon melts at 1685ÅK and boils at 3900ÅK.

The elemental silicon gives polymer in the solid state. However, silanes $H(SiH_2)_nH$ give isochains with n equal to 45. $(Si\ H)_n$ exists as high molecular mass polymer with six-membered flat rings probably as phyllo-poly (Silanes).

Phosphorus

Phosphorus in the solid state exists in three allotropic forms namely white phosphorus, red phosphorus, and black phosphorus. The solid obtained by condensing the vapour is white phosphorus which gives cubic crystals with a = 2.21Å and is known as α-phosphorus. β-phosphorus is hexagonal tetrahedral in structure composed of four atoms. Each atom occupies the comer of a regular tetrahedron.

White phosphorus upon heating or irradiation is converted to red phosphorus. Its structure is not well established. Black phosphorus is obtained by careful crystallization of white-phosphorus or subjecting it to high pressures. It gives polymeric isochain of phosphorus atoms. The orthorhombic phosphorus has a, b & c equal to 3.313Å, 10.478Å, 4.376Å respectively and Z as 8.

Sulphur

Sulphur occurs in a number of allotropic modifications of formulae $(S\text{-}S)_n$ where n can be as great as 10. Sulphur atoms are linked by covalent bonds having bond angle S-S-S as 105°. Sulphur forms isochains in the melt. Sulphur cations form S_n^{2+} (with n = 4, 8, 16) and sulphur anions form S_n^{2+} (with n = 4,9).

Sulphur in the solid form gives Sδ, Sμ, and Sn. Of these Sπ is stable. In molten sulphur, on solidiftcation, S changes to Sμ. The molecular structure of Sπ is an octatomic sulphur chain. Sλ, gives a long polymerized chain of sulphur atoms. 5A gives a crown shaped octatomic sulphur ring.

Sulphur exists in two different types of crystals namely α-sulphur and β-sulphur. α-sulphur gives rhombic or octahedral crystals, with a as 10.83A and c as 4.26A. This is stable up to 95.5°C. β-sulphur gives monoclinic or prismatic crystals with a as 8.57A, b as 13.05Å. δ-Sulphur is amorphous in nature. Other forms of sulphur are hexatomic sulphur and modifications of catena-poly sulphur.

In liquid state, the chains of sulphur are broken at higher temperature. When the molten mass is poured in cold water, it gives plastic sulphur. Sulphur above 250°C on chilling gives thin rods which can be stretched 20 times the length at the room temperature. The resulting fibre Sϕ consists of Sγ(S_8 rings) and S4 consists of long chain molecules.

IONIC CRYSTALS

We have been dealing throughout with the structures having the balls of the same size. The majority of compounds consist of different elements. We shall now try to build up structures consisting of balls of different sizes. The simplest will be those structures which have balls of two different sizes.

Radius Ratio

Radius ratios of the two different balls play an important role in deciding the shape of the ultimate structure. It is useful to consider that the smaller ball occupies holes in a closed packed lattice of the larger balls. A closed packed lattice contains triangular holes (Z = 3), tetrahedral holes (Z = 4) and octahedral holes (Z = 6). Consideration of the geometry of the holes leads to the observation that for given larger spheres of unit radius, small spheres that can fill the holes should have the radius.

Crystals

The simplest structures are of AX type, where A represents a ball of larger diameter and X represents a ball of smaller diameter. Balls with two different diameters can be arranged to give structures like AX, AX_2 and A_mX_n. Some of the important structures. The critical ratios are:

Cubic (8) 0.732 $\rightleftharpoons$ Octahedral (6) 0.414 $\rightleftharpoons$ Tetrahedral (4) 0.225 $\rightleftharpoons$ Tringular (3) 0.155 $\rightleftharpoons$ Linear (2).

(a) A face centred cubic structure or cubic close-packed structure.

(b) Three-dimensional view showing four atoms per unit cell in case of cubic close-packed structure.

(c) Monoatomic body-centred cubic structure.

(d) Three-dimensional view showing two atoms per unit cell in α-body-centred cubic structure.

TYPES OF PLASTIC STATE

The liquids on cooling, try to attain a minimum potential energy and thus, form crystal. If it so happens that after some time the process of crystallization stops and super cooling starts, we get a series of crystalline structures included in the super cooled structures. Such a state is known as plastic state. Thus, polymers whose chains are stiff at room temperature are called plastic.

Macromolecules

Macromolecules are molecules built from a large number of atoms. They can be of natural origin like cellulose, proteins, etc., or they may be produced synthetically like poly (ethylene) or silicones. All macromolecules consist of at least one chain of atoms bonded together, not necessarily by covalent bond, but sometimes by coordinate bonds or electron deficient bonds.

This is known 'as main-chain which consists of constitutional units and of at least two end groups. A polymer is a substance composed of molecules characterized by the multiple repetition of one or more species of atoms linked to each other in amounts sufficient to provide a set of properties that do not vary markedly with the addition or removal of one or a few constitutional units (IUPAC).

Constitutional Units

A constitutional unit is the smallest unit whose repetition completely describes the main chain structure. Hence, it is known as the constitutional repeating unit.

Monomeric Unit

The monomeric unit (or mer. or base unit) is the largest constitutional unit contributed by a monomer during a polymerization process.

End Groups

End groups are the groups that occur at the end of a macromolecular chain. Linear chains have two end groups but the star shaped polymer with four arms has four end groups. The end groups of a polymer yield information on the mechanism of synthesis and this information may be used to determine the molar mass or the degree of branching.

Polymer Homologous Series

If the molecules have identical constitutional repeating units but differ only in the number of such units per macromolecule, the macromolecules are known as belonging to polymer homologous series. The number of monomeric units joined together in a macromolecular chain is known as the degree of polymerization of a molecule. In exceptional cases the degree of polymerization is identical with the chain link number, which gives the number of atoms joined sequentially together in a chain.

Polymers with a small number of constitutional units and unspecified end groups are known as *oligomers*. Small refers to 2-20 constitutional units in synthetic polymers and several hundred units in nucleotide chemistry. Telomers are oligomers with end groups consisting of fragments of chain transfer agents. Telchelic polymers are oligomers with known functional end groups.

DEFECTS OR DEFORMATIONS

In actual practice it is not possible to arrange all the balls in positions required by specific structure. These balls slip past each other to produce what are known as defects. These defects may be point defects or linear deformations.

Points of Deformations

The defects may not alter the stoichiometry of the crystal, but the balls may move within the crystal itself. Normally these balls will move from an occupied position into a vacant position. These defects are called intrinsic defects. These defects are of two types. In one, the ball moves from its lattice position to an interstitial position. This type of defect consists of the vacancy plus the interstitial and is known as Frenkel defect. In the second defect a pair of balls of two different sizes leave the position in the crystal and occupy different sites. This type of defect is known as Schottky defect.

Linear Defects: Dislocations

The shifting of the balls from atomic sites in the crystal without separation of the atomic planes, will give rise to line defects, known as dislocations. Two distinct stages in the slipping from A to D and the two dislocations are known as edge and screw dislocations.

The dislocation that contains an extra half plane of atoms at the slip plane is known as edge dislocation. The line defect, from above and below the slip, generates a screw pattern. In the figure the circles are above the slip plane and the dots representing atoms are below the slip planes. The distortions may be pure edge, pure screw or any possible combination of edge and screw components.

Hetero Chains

Styrene on heating changes from a clear liquid to a solid. Berthelot called this phenomenon as polymerization. He also observed that poly (styrene) on being heated at higher temperatures converts to styrene. This leads to miscelle theory of polymers. Staudinger classified colloids into colloidal dispersions miceller colloids (associated colloids) and colloidal molecules (macromolecular).

Steward and Von Weimam suggested that colloids are general possible states of material and not specific materials. Staudinger concluded that organic colloid consists of many atoms joined together by covalent bonds, *i.e.*, true macromolecules.

STRUCTURES OF SOLIDS

The three states of matter are interconvertible. This can be done by heating or cooling. Heating increases the interparticle spacing and the kinetic energy of the particles. Many properties of solids, liquids and gases can be easily observed with the help of our sense organs. The properties which can be observed with the help of our sense organs are called macroscopic properties.

The description of the behaviour of the three states of matter an terms of atomic theory is called microscopic description of matter. In chemistry, we try to explain the macroscopic behaviour of matter in terms of its microscopic description. In this chapter, we would try to explain the observable properties of different states of matter in terms of the behaviour of the constituent particles in term.

In a solid, the constituent particles atms, ions or molecules are most closely packed and have the strongest inter molecular force of attraction. As a result particles are fixed in their position and do not have any freedom of motion except that of vibration about their mean position. This give rise to the following characteristic properties to a solid.

(i) A definite shape and definite volume

(ii) Incompressibility

(iii) Rigidity

(iv) No fluidity

(v) Poor diffusibility through them

CLASSIFICATION OF SOLIDS

Based on their structural features, the solids are classified into following three type :

(a) Crystalline solids

(b) Polycrystalline solids

(c) Amorphous solids.

These are described below.

Crystalline Solids

All crystalline solids have the following characteristics.

(i) The constituent particles in crystals are generally held by strong interatomic interionic or intermolecular forces. Particles in a solid do not possess translation motion, but they vibrate about their equilibrium positions.

(ii) Crystalline solids possess characteristic geometrical shapes.

(iii) When cut or hammered gently, crystalline solids show a clean cleavage, *i.e.*, they show fracture along a smooth surface.

(iv) Crystalline solids have sharp melting points. This indicate that there exists a long range order in crystalline solids. The long range order in crystalline solids is due to regular arrangement of the constituent particles (atoms, ions or molecules) in three dimensions.

For example in a crystal of sodium chloride, experiments (X-ray diffraction method) show that Na^+ and Cl^- ions are located at alternate sites as shown below:

Na^+	Cl^-	Na^+	Cl^-	Na^+	Cl^-
Cl^-	Na^+	Cl^-	Na^+	Cl^-	Na^+
Na^+	Cl^-	Na^+	Cl^-	Na^+	Cl^-
Cl^-	Na^+	Cl^-	Na^+	Cl^-	Na^+

The arrangement shown above is only in two dimensions. In three dimensions we find that each Na^+ ion is surrounded by a fixed number (six) of Cl^- ions and *vice-versa*. The origin of the three dimensional order in NaCl crystals is due to the strong Colombian attraction between Na^+ and Cl^- ions. A similar regular arrangement is also found in other solids.

(v) The mechanical, electrical and optical properties depend upon the direction along which they are measured, *i.e.*, crystalline solids are anisotropic in nature.

Diamond is a crystalline slid. When hammered gently, diamond get cleaved smoothly along certain preferred direction. Hammer gently a big crystal of rock salt (sodium chloride). The big crystal gets broken down into smaller crystals with smooth surfaces.

Polycrystalline Solids

In certain crystalline solids, the crystals are very find. These cannot be seen by the naked eyes. Such solids given an impression of being amorphous. Such fine crystalline solids, which appear as amorphous are termed as polycrystalline solids. Metal powders are polycrystalline in nature. In a polycrystalline sample, there are small crystals. These crystals are randomly oriented. As a result, a sample of polycrystalline material appears to be isotropic, even when each individual crystal is anisotropic.

Amorphous Solids

Amorphous solids have some mechanical properties which are commonly associated with the word solid. But, the amorphous solids differ from the crystalline solids in many respects. So, amorphous solids are sometimes called as pseudo solids.

Some characteristic properties of the amorphous solids are;

(i) The mechanical, electrical and optical properties of amorphous solids do not depend upon the direction, *i.e.*, amorphous solids are isotropic. In this respect, the amorphous solids resemble liquids. It is for this reason that the amorphous solids are considered as supercooled liquids. There is only a short-range order in amorphous solids.

(ii) Amorphous solids do not poses sharp melting points. This indicates the absence of long-range order Amorphous soften on heating and gradually begin to flow like liquids.

(iii) Amorphous solids do not occur in characteristic geometrical shapes.

(iv) When hammered, amorphous material break in an irregular manner. When you hammer a glass sheet or a plastic sheet, these get broken in a very irregular way along many directions.

Examples. Some typical amorphous materials are Plastics, Glass, Rubber, Lamp black, pitch etc.

The solid state is next to gaseous state in its simplicity to understand, although it is also not free from complex problems. Solids are obtained by cooling liquids. There are inherent two possible ways of passing from the equilibrium liquid state into solid state namely crystallization and glass transition.

Crystallization is a transition from a state of short range order to one of long range order in the process of formation of a new phase. The concept of short range order and long range order is defined by the ratio between the distance over which the order extends and the dimensions of the elements arranged in the order. This is brought about by slowly cooling the liquid where the molecules, due to enough thermal energy and low viscosity of the medium, get enough opportunity to arrange themselves in a geometric order. With the slow release of energy, they grow on the nucleus maintaining the shape of the crystal. It is a phase transition of the first order.

Glass transition does not involve a change of phase, retaining a short range order. It is actually a super-cooled liquid in a non-equilibrium state, which solidify on cooling passing into glassy state. It loses the property of liquid state and acquires the properties of solid state. These changes do not occur abruptly but gradually over a certain range of temperature.

The amorphous materials posses properties like mechanical, electrical, refractive index, etc., the same in all the directions, such materials are known as isotropic materials. The properties in a crystal depend on the direction along which they are measured. This shows an ordered arrangement of particles in a crystal, such substances are known as anisotropic materials.

A perfect crystal has a perfectly periodic structure which depends on the concentration of the solution and the rate of cooling. In actual practice, a perfect crystal is a model and not a physical reality.

CHARACTERISTICS OF SOLIDS

Some characteristic properties of solids are given as follows :

(i) Solids maintain their volume independent of the size of shape of the container in which they are placed.

(ii) Solids are rigid and have definite shape.

(iii) Solids diffuse very slowly compared to liquids or gases. This is because the constituent particles in solids are closely paced and there is very little space to permit the movement of the particles.

(iv) Solids re nearly incompressible. The compressibility of solids is about 10th times that of the gases.

(v) Most solids melt on heating. Some undergo sublimation upon beating. Melting temperatures vary considerably from solids to solid.

(vi) Solids have very high mass-to-volume ratio (density) as compared to gases.

TYPES OF FORCES

The forces that hold the atoms or molecules can be divided into three main groups:

1. Primary forces;
2. Secondary forces; and
3. Hydrogen bonding.

Primary Forces

As already discussed, it is the outermost valence electrons that play a part in bonding of atoms or molecules. These forces can be further subdivided into (a) ionic (b) covalent (c) metallic, and (d) Van der Waals. The above classification is very useful but in practice few bonds approach these types of bonds. By and large the bonds are of intermediate type as shown by Ketelaar. The triangular diagram illustrates the transition between ionic, covalent, and metallic bonding.

Ionic Forces

If a sodium atom is placed near a chlorine atom the electron transference takes place resulting in a group $Na^+ Cl^-$. If a second positive sodium ion comes near a chlorine ion, it will experience a strong attraction towards Cl^- and weak repulsion from Na^+ as the force of repulsion diminishes rapidly with increasing distance as can be seen from potential energy curve it would be possible to add another Na^+ around the Cl^-. But the sodium ions must not come too close to each other and at the same time they should not try to contact the Cl^-. This repulsion interaction between the sodium ions and stearic hinderance will not allow more than a certain number of positive ions (6 in the case of sodium) around a negative ion. Once this group of seven ions is completed, the grouping of negative ions around the external positive ions will start to restore the electrical neutrality of the aggregate.

The coulombic force between Na^+ and Cl^- is a short-range force. The homopolar bond works only in a particular direction while the ionic bond works in all the direction resulting into a space model, each ion surrounded by oppositely charged ions.

Covalent Forces

The most characteristic property in covalent bonding is the directional nature of the forces. Crystals of diamond, silicon, germanium, grey tin belong to this type of bonding. These are of two types—polar and nonpolar. The compounds containing –OH, –COOH, –Cl, –F groups are known as polar compounds like water, acetic acid, alcohol, etc. Among the most polar polymers are poly (vinyl alcohol), cellulose, starch, poly (acrylic acid), poly (methacrylic acid), poly (acrylonitrite), poly (vinylchloride), etc. The nonpolar molecules are CCl_4, benzene and polymers are polyethylene, polypropylene, polybutadiene, polyisoprene and polyisobutylene. Most of these polymers used to give amorphous structures, but with the advent of new catalysts by Nata, it has been now possible to produce the non-polar polymers as purely crystalline compounds.

Ionic Solid

In ionic solids positive and negative ions occupy the lattice points. In sodium chloride some of the lattice points are occupied by Na and the others by chloride ions. It has been observed that molten sodium chloride conducts electricity. It is reasonable to assume that the ions which produce this conductivity actually exists as charged ion in the lattice.

Further the energy required to separate the ions in the crystal lattice to an infinite distance agrees with that calculated on the assumption that the units are ions held together, by electrostatic forces. So the melting point of ionic crystals is high. (NaCl 800°, KCl 790°). Since there is no fixed directed force of attraction, they are brittle and have very little elasticity and can not be easily bent or worked. They have a great tendency to fracture by cleavage. The ions are not free to move in a solid state, hence, bad conductors of electricity but when melted becomes good conductors of electricity.

In ionic crystals, some of the atoms may be held together by covalent bonds to form ions occupying definite positions and orientations in the crystal lattice, *e.g.*, in calcium carbonate a carbonate ion does not belong to a given calcium ion but three particular oxygen atoms do belong to a given carbon atom.

CLOSEST PACKED STRUCTURE

There are many solid elements and compounds whose atomic arrangements can be visualized in terms of the closest packing of identical atoms or ions so as to use the available space most efficiently and attain maximum density.

Packing Factor

The number of atoms per unit volume is known as packing factor.

Coordination Number

The coordination number is the number of closest particles in the neighbourhood of particle in question.

Spheres in Closest Packings

A closest packing of like spheres is an arrangement of spheres which permits each sphere to be in intimate contact with the largest number of its neighbours. For spheres of equal size, this means that each sphere would have six nearest neighbours in a closest-packed layer and twelve in a three-dimensional closest packing. We shown a layer of spheres arranged in on the points of a square lattice. The efficiency of this arrangement where the area belonging to each circle is a square is calculated as follows:

$$\text{Efficiency} = 100 \times \frac{\text{Area of circle}}{\text{Area of a square}} = \frac{100\pi R^2}{4R^2} = \frac{100\,\pi}{4} \qquad ...(1)$$

Efficiency = 78.5 per cent.

The area associated with a circle here is a hexagon and each hexagon comprises of 12 right-angled triangles is $(R/2)(R/\sqrt{3}) = \frac{R^2}{2\sqrt{3}}$. The area of a hexagon is, therefore,

$$12 \times \frac{R^2}{2\sqrt{3}} = 2\sqrt{3}\ R^2. \qquad ...(2)$$

$$\text{Efficiency} = 100 \times \frac{\text{Area of circle}}{\text{Area of a square}} = \frac{100\pi R^2}{2\sqrt{3}\ R^2} = 90.7 \text{ per cent.} \qquad ...(3)$$

The packing arrangement represents the most efficient way of occupying space and this layer is called a hexagonal closest-packed layer.

In the two-dimensional hexagonal closest-packed arrangement of spheres, each sphere is surrounded by six others (nearest neighbours). Let us label each occupied site in the layer with letter. The second layer of closest-packed spheres can be added by placing spheres in the depressions or voids (or holes) of the first layer. We shall label these voids by letters B and C. In closest-packed arrangement obtained by packing spheres on B voids.

The third layer can however, be added by placing spheres on the voids marked C or the A voids created by the second layer. If we place the spheres on C, ABC sequence is obtained, if we place them on A, AB sequence is obtained. These two sequences can be repeated indefinitely to get the ABCABCABC ... and ABABAB ... arrangements; these are respectively called the cubic closest packing (CCP) and hexagonal closest packing (HCP) arrangements respectively. The same arrangement f, g, h on enlarged scale. Both these are equally efficient with 74% of available volume occupied by spheres. In both CCP and HCP structures, coordination number of a sphere remains 12. In the CCP arrangement, there is always a sphere at each face centre of a cube and CCP is, therefore, also called a face-centred cubic arrangement. Metals are generally found to crystallize with either of these arrangements; in addition, metals also crystallize in the body centred cubic structure which is not a closest-packed arrangement. This arrangement has a coordination number of 8 and only 68% of the total volume is occupied by spheres.

Voids in Closest-Packed Structures

Two kinds of voids occur in closest packings. If a triangular void in a closest-packed layer has a sphere over it we get a void with four spheres around it and such a void is called a tetrahedral void.

The B voids (over which we place the second layer spheres) are tetrahedral voids. We have depicted a tetrahedral void. The coordination number of a tetrahedral void is four. If a triangular void pointing up in one closest-packed layer is covered by a triangular void pointing down in the next layer then the void is surrounded by six spheres arranged on the comers of an octahedron such a void is called an octahedral void. The coordination number of an octahedral void is six. In C voids after the addition of the second layer represent octahedral voids.

The number of octahedral voids belonging to one sphere is given by the ratio,

$$\frac{\text{Number of otahedral voids around a sphere}}{\text{Number of spheres around a void}} = \frac{6}{6} = 1 \qquad ...(4)$$

The number of tetrahedral voids belonging to a sphere is given by the ratio

$$\frac{\text{Number of tetrahedral voids around a sphere}}{\text{Number of spheres around a void}} = \frac{8}{4} = 2 \qquad ...(5)$$

We see that in a closest-packed arrangement, the number of octahedral voids is equal to the number of spheres while the number of tetrahedral voids is double the number of spheres.

Structures of Substances Related to Closest-Packed Lattices

Besides noble gases and metals which crystallize in closest-packed structures, a large number of inorganic compounds have structures related to closest packings. In such compounds one kind of atom forms the closest-packed structure while another kind of atom goes into the voids.

In many of the solid solution of metals, similarly, some of the atoms are present in the voids. Properties of materials are often affected when foreign impurities or some of the atoms comprising the material itself are present as interstitial atoms in the voids; such interstitials are an important kind of imperfection in solids.

From the description of the closest-packed structures, we draw some inferences regarding the structures of simple compounds. For example, if in a compound of the general formula AB, the atom (or ions) form a close-packed lattice, then A atoms (or ions of opposite charge) can occupy all the octahedral voids or one-half of the tetrahedral voids. If the formula of the compound is A_2B, all the tetrahedral sites will be occupied by A atoms. Typical descriptions of ionic solids forming nearly close-packed structures are given below:

It must be remembered that the above descriptions are not exact in the sense that some of these ionic compounds are not strictly the closest-packed structures. For example, in NaCI the ions along the diagonal of the cube face do not touch each other. An interesting case of packing is given by the oxide of iron, FeO, Fe_2O_3 and Fe_3O_4. FeO has the rock-salt (NaCI) structure and ideally Fe^{+2} ions should occupy the octahedral voids. However, this oxide is always non-stoichiometric and the composition is generally $Fe_{0.95}O$. We can obtain this composition if a small number of Fe^{+2} ions are replaced by Fe^{+3} ions in the octahedral voids.

If we convert two-thirds of the Fe^{+2} ions into Fe^{+3} ions, we get the composition, $FeO.Fe_2O_3$ or Fe_3O_4; in Fe_3O_4, all Fe^{+2} ions are in octahedral voids and the Fe^{+3} ions are equally distributed between octahedral and tetrahedral voids. This is the well-known inverse spiral structure; another example of this structure is $MgFe_2O_4$ where Mg^{+2} replace Fe^{+2} of Fe_3O_4. In the normal spiral structure compounds of the general formula AB_2O_4 (*e.g.*, ferrites like $ZnFe_2O_4$; $MgAl_2O_4$), the divalent A^{+2} ions are in tetrahedral voids and the trivalent B^{+3} ions are in octahedral voids.

We see from the earlier discussion that ionic sizes determine the nature of packing in the close-packed structure. Thus, in ZnS the small Zn^{+2} ions occupy the voids while the large S^{-2} ions form the close-packed lattice. In NaCI, on the other hand (where the ionic sizes are similar), Na^+ ions occupy the octahedral voids. The coordination number of an ion (or an atom) in the close-packed structures depends on ratio of its radius with respect to the radius of the atoms surrounding it. In the case of ionic solids, this radius ratio is the ratio of the cation radius to the anion radius.

Generally, in ionic solids, cations tend to get surrounded by the largest possible number of anions and therefore, larger the radius ratio larger is the coordination number of the cation. We have summarized below the range of radius ratios over which coordination numbers of 2, 3, 4, 6, 8 and 12 are likely to be stable.

By employing value of ionic radii for cations and anions, it is possible to predict the coordination number of cations in compounds. Radius ratio is indeed a useful concept for prediction of structures of ionic compounds.

$$\text{Cubic (8)} \overset{0.732}{\rightleftharpoons} \text{octahedral (6)} \overset{0.414}{\rightleftharpoons} \text{tetrahedral} \overset{0.225}{\rightleftharpoons} \text{triangularic} \overset{0.155}{\rightleftharpoons} \text{linear}$$

Metallic Bonding

Nearly eighty elements exists as metals at NTP. Metals differ from other substances in terms of their electronic configuration. Metals interact by sharing their electrons not between its two atoms but amongst all the atoms. Thus, a metal may be viewed as a collection of positive ions embedded in a sea of mobile electrons. Such a cloud of electrons often referred to as electron gas is responsible for most of their electrical properties.

Other Types of Bonding

The classification of solids into four major types should not be taken in simply that there is a sharp demarcation between one type and another. In some solids, more than one type of bonding are responsible for the stability of structures.

As mentioned earlier. Van der Waals forces are responsible for inter-layer binding in graphite. Similarly, there are susbtances which form thread-like structures (*e.g.*, amorphous sulphur, asbestos) where atoms are linked to one another along a chain by covalent bonds, but the chains are held together by weak Van der Waals forces. In such substances, threads (or fibres) can be pulled out easily, but it is relatively harder to break threads. There are solids like the metal silicates where the bonding is between ionic and covalent bonds. Silicates form thread-type, sheet-type or three-dimensional structures, held together by ionic interaction with cations present in them.

There are solids like ice which are held together by weak hydrogen bridges or hydrogen bonds formed between molecules (H_2O molecules in the case of ice). In proteins, stable chains of atoms are formed by covalent as well as hydrogen bonds but in the solid state the chains are held together by Van der Waals forces between them.

IONIC CRYSTALS

Solids in which ions are arranged in some definite geometrical pattern are called ionic crystals. These ions are bound by plane surfaces called faces. These faces intersect at an angle giving definite shape to the crystal. Steno observed that the shape of a crystal depends on the details of crystalization but at a given temperature the angle between planes of a crystal always remain constant. This is known as Stenos law. That is why a crystal on cleavage maintains its geometry. Sometimes a substance crystallizes out in different shapes. This is known as polymorphism.

Laws of Rotational Indices

The law of rotational indices states that the intercepts of the plane of various faces of crystal on a suitable set of axes can be expressed by small integral multiples of three unit distances.

Space Lattices

The diffraction of X-rays by crystals is a phenomenon of great importance. Diffraction studies provide much of the structural information on the arrangement of atoms, ions or molecules in crystals. Hauy suggested that the environment about any particle in space, *i.e.*, in three direction is the same as that about any other particle. Thus, a regular three-dimensional array of points, representing particles, is known as space lattice.

Unit Cell

A unit cell in the crystal is the smallest repeating unit in three dimensions, which reproduces the whole crystal by the displacement of the unit along the direction of the three axes. The crystal may be considered to consist of a very large number of unit cells, each one in direct contact with its nearest neighbours and all are similarly oriented in space. The shape and size of a unit cell are given by the spacings along the three axes (a, b, c) and the three angles (α, β and γ) between the faces.

Zero Point Energy

A crystal is found to have some vibrational and translational energy at absolute zero. This is known as the vibrational energy at absolute zero or the zero point energy of the crystal.

Indices

The intercepts on the axes of X, Y, Z of a unit cell are a, b, c, and are known as Weiss indices. The reciprocal intercept $\frac{a}{oA}$; $\frac{b}{oB}$ and $\frac{c}{oC}$, *i.e.*, h, k, I are known as Miller indices. They are always integral.

Weiss indices	2a,	b,	l/2c,	a,	b,	c,	a,	1,	1
Miller indices	1,	2,	4,	1,	1,	0,	1,	0,	0

Depending upon the magnitude of Weiss indices and the angles between them, there are seven different types of crystal system.

Bravais Lattices

In 1848, Bravais showed that there are fourteen ways of arranging points space so that each point has identical surroundings. These lattices are known as Bravais lattices.

Symmetry Properties

The arrangement of crystal faces such that the crystal transform into image of itself is known as symmetry operation. These operations are as a result of some symmetry elements. These symmetry elements are international symbols. An axis of rotatory inversion is the combination of a rotation axis with a centre of symmetry. It is denoted by placing a bar over the symbol. The various symmetry elements.

There are 32 possible combinations of these symmetry elements. These define 32 crystallographic point groups which determine the 32 crystal classes. The symmetry operations are reversible, *i.e.*, A(BC) = (AB)C. Only holohedral class of crystals follow the complete symmetry. Crystalline rocksalt and iron pyrites are cubic crystal. Rocksalt crystal possess the full symmetry of the cube but iron pyrites does not show all the symmetries because all the faces in iron pyrites are not equivalent.

Space Groups

The classification of crystals is based on the various groups of symmetry elements operations, such that they leave atleast one point in the crystal as invariant. Hence, they are called point groups. The possible three dimensional lattices symmetry operations of infinite figures are called space groups. It has been worked out that the internal symmetry of crystal may be classified according to 230 space groups, *i.e.*, there are only 230 ways of repeating a particular pattern.

Table 3

No symmetry	1	Four-fold rotation axis	4
Mirror plane of symmetry	m	Four-fold rotatory inverter	4
Two-fold rotation axis	2	Six-fold rotation axis	6
Three-fold rotation axis	3	Six-fold rotatory inverter	6
Three-fold rotatory inverter	3	Centre of symmetry inverter	1

Diffraction Methods

In 1912 Von Lave suggested that a crystal can act as a three-dimensional diffraction grating for X-rays of wave length ranging 0.001-50Å. X-rays are produced by striking atoms in a solid target by accelerated electrons. Thus, the electrons in the inner orbit of the target are displaced by bombardment and the energy liberated by the transition of electrons from a higher energy level to the vacancy in the inner orbit is given out in the form of an x-ray photon.

These transitions produce characteristic lines in the x-ray spectrum. A copper target gives an intense line at 1.54Å. In addition to the above radiation, a continuous range of X-ray wave lengths are obtained. These are called white X-rays. Thus, using a nickel filter radiations of wave length 1.54Å can be picked up which are widely used for polymers.

Beams of electrons and neutrons has also been diffracted by crystals. Neutron diffraction determines the position of protons in a structure. X-rays scattered by electrons do not generally lead to a precise position of protons. Electrons do not penetrate very far into solids, but their use in gases has been found very valuable in finding the structure of gas molecules. Reflections from the surface reveal facts about their structure.

Hugens-Fresnel suggested that every point reached by the electromagnetic vibrations may be regarded as a radiation centre. Bragg suggested that it would be convenient to assume that the X-rays are reflected from the planes in the crystal. Woulfe-Bragg found a simple relation between A the wave length of the X-rays, d the distance between planes in a crystal and Q the angle of reflection

$$n\lambda = 2\ d\ \sin\theta \qquad \text{...(6)}$$

Let the plane ABC be perpendicular to the incident beam of parallel X-rays and the plane DEF is perpendicular to the reflected rays. It can be seen that the ray BPE travels farther from AOD by a distance equal to LP + PM. The angle of reflection is equal to the angle of incidence θ and LP & PM is equal to the integral number of wave lengths.

$$\therefore \quad LP + PM = n\lambda \qquad \text{...(7)}$$

$$LP = PM = d\ \sin\theta \qquad \text{...(8)}$$

$$\therefore \quad n\ \lambda = 2d\ \sin\theta \qquad \text{...(9)}$$

When n = 1, it is called the first order reflection and n = 2, it is called the second order reflection, etc. Equation 8.4 can be written as

$$\lambda = 2\left(\frac{d}{n}\right)\sin\theta \qquad \text{...(10)}$$

$$= 2\ d_{hkl}\ \sin\theta$$

where d is the distance between the planes having hkl as miller indices.

The second order reflections from 100, 110, 111 planes are labelled as 200, 220, 222 respectively. In a cubic crystal distance between planes with Miller indices hk*l* is given by

$$d_{khl} = \frac{a}{(h^2 + k^2 + l^2)^{1/2}}$$

when a is the length of the side of the unit cell.

Thus, the spacings between the planes in a primitive cubic crystals may have values a, $\frac{a}{\sqrt{2}}, \frac{a}{\sqrt{3}}, \frac{a}{\sqrt{2}}, \frac{a}{\sqrt{5}}, \frac{a}{\sqrt{6}}, \frac{a}{\sqrt{8}}$ but not $\frac{a}{\sqrt{7}}$ as $h^2\ k^2 + l^2$ cannot have a value.

In the case of face centred cubic lattice, the spacings which will appear are $\frac{a}{\sqrt{3}}, \frac{a}{\sqrt{4}}, \frac{a}{\sqrt{8}}, \frac{a}{\sqrt{11}}, \frac{a}{\sqrt{12}}, \frac{a}{\sqrt{16}}, \frac{a}{\sqrt{19}}, \frac{a}{\sqrt{20}}$, etc.

In the case of body centred cubic lattice the inter planer spacings are $\frac{a}{\sqrt{2}}, \frac{a}{\sqrt{4}}, \frac{a}{\sqrt{6}}, \frac{a}{\sqrt{8}}$, etc.

Thus, knowing the missing X-ray lines it would be possible to predict the type of cubic lattice of a given crystal.

If X is known and the direction of θ in which the intensity of the scattered radiations is maximum is determined experimentally, d can be calculated and a can be known from the density of the cubic crystal.

Example:

The length of the side of a unit cell of potassium is 5.34Å. Calculate the distance between 200, 110 and 222 planes.

$$d = \frac{5.34}{\left(h^2 + k^2 + l^2\right)^{1/2}}$$

For 222 planes d is $\frac{5.34}{(4)^{1/2}} = 2.67\text{Å}$

For 110 planes d is $\frac{5.34}{(2)^{1/2}} = 3.77\text{Å}$

For 222 planes d is $\frac{5.34}{(12)^{1/2}} = 1.54\text{Å}$

Macromolecules : In macromolecules, there are three basic types of ordering; the small period, the helicals when repeat distance is along the chain and the large period.

The small period characterizes the dimensions of the smallest structural elements out of which all polycrystalline bodies are made in their various combinations.

Regularly built macromolecular chains with side substitutes assume in space helical conformation with constant pitch. The helical pitch conforms to the periodicity along the chain with a specific size of the order of several angstroms.

A special type of ordering inherent only to polymer bodies is characterized by the alternate regions of long-range and short-range orders and in the limit; amorphous and crystalline regions are obtained in oriented polymers. The large periods have the order of hundred of angstroms. Three methods are used with crystals.

Laue's Method

White x-rays are used in this method. The x-ray pattern obtained represents a system of spots from various systems of planes in a crystal. This is not used for polymers because polymer single crystals of large size are usually not obtained.

Bragg's Method

In this method a monochromatic radiation is used and the sample is rotated or tilted when it is photographed. The photographic film is arranged on the surface of the cylinder around a sample which is placed along the cylinder axis. Diffraction maxima appear in the coarse of rotation and then θ can be measured.

Debye Scherrer Method

Samples in the form of powder or a polycrystalline body are examined. The incident monochromatic X-ray is diffracted on the planes of a crystallite of a polycrystalline sample that satisfy the Woulfe-Bragg equation by their orientation with respect to incident ray.

Diffracted rays will spread along a cone with the vertex angle equal to 2θ. The intersection of these cones with the flat photographic film placed perpendicular to the incident ray behind the sample produces a system of concentric rings of growing radii, each of which contains all the reflections with the same angle θ.

The reflections are recorded on circular photographic film. If coarse crystals are used, the powder pattern produces rings of spots, each spot produced by the different crystal planes. If the crystals are very fine, continuous arcs are obtained on the film.

X-ray diffraction analysis is used in obtaining the structure of polymeric bodies and its change as a result of heat, mechanical and other effects, on the phase transformation of a body and to determine the conformation of chains. They also give average dimensions of crystal and their distribution according to dimensions. They make it possible to detect the origination of the smallest (10-100 A) cracks in polymeric bodies.

Primitive Cubic Lattice

In a primitive or simple cubic cell, the cations are located only at the corners of the unit cell and its centres occupy the anions, *e.g.*, Cs Cl, *i.e.*, it contains one ion pair per unit cell. Each cation is therefore, surrounded by eight anions and vice versa. This arrangement is known as cubic coordination. This type of structure is exhibited by cesium and thalium chloride, bromide and iodides and by a host of other compounds. The packing factor is 0.52.

$$P.F. = \frac{\frac{4}{3}\pi r^3}{a^3} = \frac{4\pi}{3}\frac{(r)^3}{8r^3} = 0.52 \qquad ...(11)$$

Face Centred Cubic Lattice

This is known as rock salt structure. There are four ion pairs or molecules in a unit cell of sodium chloride. The ions at the corners are shared by eight unit cells while those on the edges by four unit cells.

Considering the pattern made by cations, it is found they occupy all the corners of a cell and centres of the faces of the unit cell. Therefore, it is called a face centred unit cell. Similarly, the anions will also have a face centred unit cell. It has a octahedral coordination with a total coordination number of 12, 6 for the cation and 6 for the anion. So each ion is surrounded by six other ions.

This is a very common structure and is assumed by alkali halides, most of oxides, sulphides, selenides and tellurides of the alkaline earth metals. Packing factor (P.F.) is given

$$P.F. = \frac{4}{3}\pi r^3 \times \frac{4}{a^3} = \frac{\pi r^3}{\left(2\sqrt{2}\right)^3 r^3} = 0.74 \qquad ...(12)$$

Body Centred Cubic Lattice

Most of the metals show a body centred crystalline lattice Fig. 2.16c & d. The alkali metals, barium, iron, chromium, molybdenum are some of the examples. There are eight atoms at the corners of the unit cell and one is at the centre. Hence, the coordination number is eight. Goldschmidt derived an emperical ratio between the metallic radius and the coordination number.

Coordination number	12	8	6
Atomic radius ratio	1.00	0.97	0.96

$$P.F. = \frac{4}{3}\pi r^3 \times 2a^3$$

$$= \frac{8}{3}\pi r^3 \times \frac{3\sqrt{3}}{64\,r^3} = 0.68 \qquad ...(13)$$

The structure of titanium (and arsenic). It is composed of octahedra each with a titanium ion at the centre and the octahedra shares edges and corners with each other.

The structure of arsenic. It is a NaCl type of cell with all the atoms of the same kind. It is deformed along a three-fold axis which gives a corrugated layer structure formed by pyramidal groups of atoms.

Fluorite Structure

In the fluorite each cation is surrounded by eight anions and each anion is surrounded by four cations to give compounds of the type AX_2. The anions have a tetrahedral coordination with coordination number 4, while cations are surrounded by eight anions at the corner of a cube giving a cubic coordinations. The overall structure is known as 8:4 coordination. The compound will have a formula MX_2.

Calcium fluoride, and difluorides of strontium and barium, etc. give fluorite structure. Their radius ratio lies between 0.73 to 1.00. The oxides, sulphides, selenides, tellurides of lithium, sodium, potassium give antifluorite structure which is the same with positions of cations and anion interchanged. The formula of their compounds will be M_2X.

The Rutile Structure

In this structure the unit cell is tetragonal where a = b but not equal to c. It is exhibited by compounds like titanium dioxide. The six oxygen ions are at the corners of the octahedron and the titanium ion at the centre of it. Three titanium ions surrounding each oxygen ion are coplaner. Hence, it gives an overall coordination of 6 : 3. When these octahedra are joined together each anion forms part of three octahedra. The formula is M_2O_4 or MO_2 and the radius ratio will be greater than 0.414. Oxides of tin, lead, etc., and difluorides of transitional metals give rutile structure on crystallization.

Zinc-Blend Structure

Another important structure is zinc sulphide lattice. Zinc sulphide crystallizes in two structures, sphalerite lattice and wurtzite lattice. The wurtzite structure differs from sphalerite in being tetragonal rather than cubic. Sulphide ions form a face centred cubic lattice with zinc ions occupying alternate tetrahedral sites.

Both zinc and sulphur have coordination number four and the radius ratio is less than 0.414. Zinc sulphide, halides of copper, silver iodide, beryllium sulphide and calcium fluoride crystal have zinc blende structure. They form 1 : 1 compound in which cation is much smaller than the anion.

CRYSTAL STRUCTURE

Bragg during his study of X-ray diffraction derived a relation between the distance between planes in a crystal and the angle at which the reflected radiations have a maximum intensity.

$$\lambda = 2\left(\frac{d}{n}\right)\sin\theta = 2\, d_{hukl}\sin\theta \qquad ...(14)$$

where n is the order of reflection usually set equal to 1, d is the distance between planes, θ is the angle and λ is the wave length employed.

In a cybic system $$d_{hkl} = \frac{a}{\left(h^2 + k^2 + l^2\right)^{1/2}} \qquad ...(15)$$

$$= \frac{2a\sin\theta}{\left(h^2 + k'' + l^2\right)^{1/2}} \qquad ...(16)$$

thus, knowing h, k, *l*, θ and d_{hkl}, a can be calculated.

The Cohesive Energy of Crystals

Born Mayer derived the equation for crystal energy per mole

$$U = \Delta H_{cryst} = \frac{Z+Z-C^2NA}{r_o}\left(1 - \frac{\rho}{r_o}\right) \qquad ...(17)$$

$$= \frac{Z^2\varepsilon^2 A}{r_o}\left(1 - \frac{\rho}{r_o}\right)$$

where ZC is the net charge on ions, r_0 is the equilibrium inter nuclear separation, A the Madelung's constant, and p the repulsion energy constant and is expressed in terms of volume compressibility β. It has an average value of 0.311×10^{-8} cm for alkali halides.

In the case of NaCl the value of r_0 is 2.81 A,

A is 1.75 and ρ is 0.311×10^{-8}cm

$$\Delta H_{NaCl} = U_{NaCl} = -\frac{(4.8 \times 10^{-10})^2 \times 6.02 \times 10^{23} \times 1.75}{2.81 \times 10^{-8}\ cm}$$

$$\left(1 - \frac{0.311 \times 10^{-8}\ cm}{2.81 \times 10^{-8}\ cm}\right)$$

$$= 7.68 \times 10^{12}\ crg/mole$$

$$= -184\ K\ cal/mole$$

which is in good agreement with the experimental value of –183.3 Kcal/ mole.

Compressibility

The compressibility β is given by Born and Landes equation

$$\frac{1}{\beta} = -\frac{mnu_o}{9\upsilon_o} = -\frac{nu_o}{18r_o^3} \qquad ...(18)$$

$$\therefore \qquad n = 1 + \frac{18r_o}{\beta AZ^2\varepsilon^2} \qquad ...(19)$$

The lowest potential energy per molecule is $U_o = \frac{AZ^2\ \varepsilon^2}{r_o}\left(1 - \frac{1}{n}\right)$

where n is the repulsive force constant ε the charge, U_o is molecular volume equal to $2r_o^3$.

Born-Haber Cycle

Bom-Haber made the direct determination of the crystal energies. In the case of sodium chloride

$$\begin{array}{ccccc} Na^+(g) & + & Cl^-(g) & \xrightarrow{\Delta H^o_{Cyst}\,(NaCl)} & NaCl\ (cryst) \\ & -\Delta H^o_{ion}(Na,\ g)\downarrow & & \downarrow\Delta H^o_{ion}(Cl^-,\ g) & \\ & \downarrow\Delta H^o_f(NaCl) & & & \\ Na(g) & + & Cl(g) & \xrightarrow[-\Delta H^o_f(Cl,g)-\Delta H^o_f(Na,g)]{} & Na(s) + 1/2\ Cl_2(g) \end{array}$$

$$\therefore\ \Delta U = \Delta H^o_{cryst}\ (NaCl) = \Delta^o_f H_f\ NaCl\ (cryst) - \Delta H^o_f Na(g) - \Delta H_{ion}\ Na^+(g)$$

$$-\Delta H^o_f Cl(g) + \Delta H^o_{ion} Cl^-(g)$$

$$= (-98.23 - 25.9 - 119.9 - 29.01 + 86.8)\ Kcal$$

$$= -186.2 Kcal$$

Δ^o_f NaCl (cryst) = the crystal energy

Δ^o_f NaCl = standard heat of formation of NaCI (s)

Δ^o_f Na(g) = heat of sublimation of Na (s)

Δ^o_f Cl(g) = energy of dissociation of Cl_2 (g) into atoms

Δ^o_{ion} Na^+ (g) = Ionization potential of Na

Δ^o_{ion} Cl^-(g) = electron affinity of Cl

Structure of Simple Ionic Compounds

It is much more difficult to describe the crystal structures or unit cells of compounds compared to those of elements. In the case of an element like copper (FCC), all the lattice points are occupied by copper atoms.

However, if we were to consider the structure of the simplest ionic compound of the formula AB, we will have to describe the arrangement of both A and B ions. We shall now briefly examine the structures of a few simple ionic compounds of the type AB and AB_2.

Compounds of the type AB are generally cubic and can possess one of the three types of structures: cesium chloride (CsCI) type, rock salt (NaCI) type and zinc blende (ZnS) type.

Chlorides, bromides and iodides of lithium, sodium, potassium and rubidium (as well as some of the halides of silver), all of which ordinarily possess the NaCl structure with 6:6 coordination, transform to the CsCl structure with 8 : 8 coordination on application of pressure; notice that high pressure increases the coordination.

On the other hand, CsCI on heating transforms to the NaCl structure at 490°C; similarly, rH_4Cl, NH_4Br and NH_4I transform from the CsCl structure to the NaCI structure at 184°, 138° and –18°C respectively.

Table 4

	Description and examples	*Coordination number*	*Number of molecules per unit cell*
CsCl-type	Cs^+ at the body centre of a cube of Cl^- and Cl^- at the centre of a cube of Cs^+, *e.g.*, CsCl, CsBr; Csl, TIC1.	Cs^+8 Cl^-8	1
Nad-type	Both Na^+ and Cl^- form FCC lattices and the two interpenetrate each halfway, *e.g.*, All sodium halides, AgCl, Transition metal monoxides.	Na^+ 6 Cl^- 6	4
ZnS-type	This also has a face centred arrangement of ions, but more complex than NaCI, ZnS, CuCl.	Zn^{+2} 4 S^{-2} 4	4

Many of the compounds of the formula AB_2 crystallize in the cubic fluorite (CaF_2) type structure or the tetragonal rutile (TiO_2) structure. In the CaF_2 structure, each F" ion is coordinated by four Ca^{2+} ions and each Ca^{+2} ion by eight F^- ions (cubic coordination). Other examples of compounds with CaF_2 structure are SrF_2, CdF_2; PrO_2 and ThO_2. In the rutile structure, the titanium atoms form a body-centred tetragonal structure and each titanium is surrounded by three titanium ions 3 : 6 coordination). Typical examples of rutile phases are RuO_2, SnO_2, PbO_2 and MgF_2.

The structures of many of the ionic compounds can be understood in terms of the packing arrangements of ions (and the relative sizes of cations and anions).

Ionic Radii

In the absolute sense of the word it is not a constant parameter. It is not possible to find the exact ionic radii as ions are not rigid bodies, but for all practical purposes if we take half the distance of approach between two similar ions, as the ionic radii, it will give a very useful data to build up the concept of the ionic structure in the compounds. The sum of the ionic radii of both ions in compound gives idea of the approach distance between two ions, which in turn gives the idea of their geometrical arrangement. The radius of an ion remains the same in various crystals as the repulsive force increases very sharply as the internuclear distance becomes smaller than a certain value.

Four anions are arranged around a cation. This is the limiting condition of stability as the anions are touching the cations.

$$XY^2 + YZ^2 = XZ^2 \qquad ...(20)$$

$$(2r_a)^2 + (2r_a)^2 = (2r_a + 2r_c)^2$$

or

$$r_a + r_c = \sqrt{2}\, r_a \qquad ...(21)$$

or

$$r_c = r_a(\sqrt{2} - 1) = r_a . 414$$

$$\frac{r_c}{r_a} = 0.414 \qquad ...(22)$$

In a three-dimensional arrangement it will represent an octahedral arrangement with two additional ions one above and one below the cation. By increasing or decreasing the radius ratio various arrangements are obtained. If the ratio is decreased, the anions would tend to arrange tetrahedrally. If the ratio is increased above 0.73 a cubic arrangement is obtained, *e.g.*, CsCl.

Similarly, in the case of a tetrahedral the maximum radius ratio will be 0.225. Triangular and tetrahedral arrangements

In DOAC, $(r_a + r_c)^2 = \frac{3a^2}{4}$ (body diagonal)

$$a^2 = \frac{4}{3}(r_a + r_c)^2$$

From ABC, $a^2 + a^2 = (2\,r_a)^2$ (face diagonal)

$$a^2 = 2r_a^2$$

$$2r_a^2 = \frac{4}{3}(r_a + r_c)^2$$

$$r_a + r_c = \sqrt{\frac{3}{2}}\, r_a$$

$$\frac{r_c}{r_a} = \sqrt{\frac{3}{2}} - 1$$

$$= 1.225 - 1 = 0.225.$$

Generally in ionic solids, cations tend to get surrounded by the largest possible number of anions and therefore, larger the radius ratio larger is the coordination number of the cation. By employing value of ionic radii for cations and anions, it is possible to predict the coordination number of cations in compounds. Radius ratio is indeed a useful concept for prediction of structures of ionic compounds.

Properties of Ionic Solids

The ionic bond is a strong bond, its cohesive energy is large. It is difficult to separate the ions from the condensed aggregate. Their melting points and boiling points are high but have low vapour pressure.

There is rupture of bonds on melting although, the ions are grouped together even in melt so much so that they have strong ionic interaction. The structure of solid is very compact as the ions are closely packed.

MOLECULAR CRYSTALS

Substances that exist as aggregates of discrete molecules in the crystalline state are known as molecular crystals. Their melting points are low and are efficiently packed. They are held by Van der Waals forces.

They are either cubic close packed or hexagonally close packed. The interstitial sites or holes are larger in the case of HCP than C.C.P. The tetrahedral hole can accommodate a sphere of radius 0.41 times that of larger sphere while the tetrahedral hole can accommodate a sphere of radius 0.23 times that of larger sphere.

The solid inert gases being spherical in shape form a simple CCP structure. While N_2 and O_2 which are non-sphereoids give a face centred packing CCP. In such crystals there are two types of interatomic distances, one between the atom and its molecular partner and the other between atoms not covalently bonded to one another. These are known as Van der Waal radii. In iodine the internuclear distance is 2.68Å, while the distance between two non-bonded atoms is 4.54Å. The Van der waals radius of iodine if about 2.3Å.

Table 5

Element Van der Waals	H	N	O	F	P	S	C	I	As	Se	Br	Sh	Te
Ratio A	1.20	1.50	1.40	1.35	1.90	1.85	1.80	2.00	2.00	1.95	2.2	2.20	2.15

Sulphur, selenium and tellurium have similar structure of long spiral chain packed together with their axis parallel. Phosphorous, arsenic and antimony are dimorphic. They assume both close packed tetrahedral structure and layer type of arrangement.

In BCl_3 or BF_3, the halogen atoms form an HCP structure with boron atoms in the holes.

The cohesion energy of a molecular crystal is defined as the enthalpy of the reaction

$$M(g) \xrightarrow{\Delta H} M\ (\text{Crystal}) \qquad ...(23)$$

$$\text{or} \quad (H_{fus} + H_{vap}). \qquad ...(24)$$

If the close-packed molecules possess a dipole moment then dipole-dipole interaction will take place and this, will contribute to the cohesion energy. Table 5 gives various physical properties. Special properties in the case of H_2O, NH_3 and HF are found due to hydrogen bonding. Hydrogen bonded chains in solid methanol and HF.

In the case of those molecules which are non-spherical, like organic compounds, they can still be packed in a compact way. Solid ethane exhibits a cubic structure but solid ethane and ethylene have HCP with body centred structure. In the case of aromatic compounds some times the planes of molecules are parallel to the crystal axis, *e.g.*, benzene.

In the case of naphthalene and phenanthrene, anthracene is better packed than phenanthrene as can be seen from their melting and boiling points.

Table 6 : Physical properties of molecules or ions in molecular crystals.

Molecule or ions	*mp* °*K*	*bp* °*K*	*ΔH fus* *K cal/mole*	*ΔH vap* *K cal/mole*	*μ* *ion debyes*	*Polarizability* *α cm³ × 10⁻²⁵*
He	3.5	4.2	0.005	0.02	–	2.0
H_2	55.2	85.2	0.37	1.51	–	–
F	–	–	–	–	–	–
Cl_2	172.0	239.0	1.53	1.53	–	46.1
Br_2	266.0	298.0	2.52	2.52	–	–
Br	–	–	–	–	–	–
H_2	14.0	20.4	0.03	0.22	–	7.9
W_2	63.1	77.3	0.17	1.33	–	17.6
O_2	54.0	90.2	0.11	1.63	–	16.0
S_8	368.0	–	–	3.01	–	–
HF	190.0	190.0	0.09	–	1.98	24.6
HC1	158.0	158.0	0.48	3.86	1.08	26.3
HBr	186.0	186.0	0.56	4.21	0.79	36.1
H_2O	273.0	373.0	1.44	9.72	1.84	–
NH_3	195.0	240.0	1.35	5.58	1.45	–
SO_2	198.0	263.0	1.77	5.96	1.62	–
CO_2	195.0	–	–	6.03	–	26.5
BF_3	145.0	–	–	5.7	–	–
SiF_4	177.0	–	–	6.15	–	–
CH_4	90.7	112.0	0.23	1.96	–	–
SiH_4	89.0	161.0	0.16	2.9	–	–
HCOOH	282.0	373.0	3.03	5.32	1.2	–
CH_3COOH	290.0	391.0	2.80	5.83	1.86	–
CH_3OH	175.0	338.0	0.76	8.94	1.68	–
C_2H_5OH	159.0	352.0	1.20	9.22	1.70	–

Table 7

	Mol wt	*Molar volume*	*mp °C*	*bp°C*
Anthracene	178	142	217	354
Phenanthrene	178	174	996	340

Naphthalene is soluble in benzene but insoluble in water. Anthracene is less soluble than phenanthrene. They are bad conductors of electricity as they do not have free electrons or ions.

SULPHUR

Sulphur acts as a diatomic molecule only at high temperatures in gaseous state. These vapours on cooling give a ring structure of S_8. It gives two structures rhombic sulphur and β monoclinic sulphur. Rhombic sulphur is stable upto 89° and has a unit cell of 16 S_8 molecules in a ring form with S-S bond length 2.10A and SSS as 103°. β-monoclinic sulphur is stable from 98° to 122°. Sulphur on heating gives liquid which becomes viscous as the temperature is raised to 160°, but gives an amorphous plastic mass on cooling rapidly. This is due to large interaction between sulphur molecules. X-ray studies indicate that it has a helical structure with eight atoms per spiral. The S_8 ring opens and joins to form a chain resulting in a polymer with very small S_8 molecules. Sulphur has been reported to exist in fifteen forms. The solubility of sulphur depends on the arrangement of molecules, CS_2 only dissolves molecules like S_8 whereas long chain molecules are insoluble.

COVALENT SOLIDS

The atoms in a covalent solid occupy such position that the attractive force due to two protons acting on each electron completely counter balances the force of repulsion between two electrons. They equally share electrons with their neighbour. These bonds extend in fixed directions resulting into a giant interlocking structure, *e.g.*, diamond, silicon, germanium, grey tin, etc.

The covalent crystals have high energies (35-150 Kcal/mole), short interatomic distances (1.01-1.6 A), relatively constant angles between successive bonds. These bonds are very strong (high melting point and large heat of sublimation). Their directional character does not lead to maximum filling of space. They are bad conductors of electricity and are chemically inert.

The bonding in these crystals is through sp^3 hybridization. Diamond is insulator, silicon, germanium and grey tin are semiconductors while white tin is metallic.

Carbon also assumes graphite structure. The bonding in graphite is with sp^2 hybrid orbitals forming a bonds linking the carbon atoms. The MO approach considers the crystal as a large aromatic molecule with π MO'S delocalized parallel to the layers.

Internal Symmetry

Wells has shown that the symmetrical arrangements of atoms around a given atom is indicative of the type offers governing their interaction. A compound containing A and B have stoichiometric ratio AB_3.

The arrangement given in (a) indicates that it is a molecular compound (Fe_3) while the arrangement (b) shows the presence of AB_2 & AB_4 groups, *i.e.*, the number of bonds in which A is engaged with B is either two or four ($AuCl^-_3$). The arrangement in C indicates that the compound exist as a dimer ($AlCl_3$). While d is a cyclic trimer (SO_3 or S_3 O_9) c & f represent a chain structure.

All these indicate that a detailed consideration of the spatial arrangement of a system can give a powerful insight of the nature of bonding. Cu, NaCl and diamond all belong to cubic system but NaCl has a heteronuclear linking and in contrast to Cu, diamond has a tetrahedral structure. These differences can only be explained by considering internal symmetry and space groups.

METALS

Those crystalline solids which exhibit high thermal and electrical conductivity, a metallic lusture, a large coefficient of thermal expansion and low ionization energy are known as metals. The structure and properties of metals are given in Table 8.

Metallic elements display body-centred, face centred or HCP structures. In metals the radii are found to be functions of coordination number and very often all the neighbours are not equidistant. Goldschmidt deduced that the relative order for different coordination numbers are:

Hume-Rothery have shown that metals will form alloys provided the following conditions are satisfied.

Table 8 : Structures of some polymorphic metals

	Structure	*Se* *angstroms*	*Temperature* °*K*
α Ca	CCP	3.94	291
γ Ca	BCC	3.87	770
α Fe	BCC	2.48	293
α Fe	CCP	2.57	1189
δ Fe	BCC	2.53	1667
α Sr	CCP	4.29	298
β Sr	HCP	4.31	521
γ Sr	BCC	4.19	887
Li	BCC	3.03	293
Li HCP 3.11 351			

1. The metallic radii should differ less than 15%.
2. The crystal structures of the metals should be identical.
3. The metals must be chemically similar.

Thus, copper and zinc form alloys with five different structures.

They also form *interstitial compounds* with non-metals like carbide, boride and nitride which have a very high melting point and extreme hardness. They also form hydrides with CCP structure. They are regarded as close-packed structures of metals with smaller non-metallic elements located in interstitial sites or holes which are either octahedral .or tetrahedral.

Let us examine the case of lithium on the basis of atomic orbital (a.o.) and molecular orbital (m.o.) bonding.

There are as many m.o.'s as there are a.o's (with each m.o. holding two electrons with opposite spins) and since 2s a.o. of each lithium atom has only one electron, it follows that half the m.o.'s are fully occupied and the other half are vacant. In other words, the 2s-band in lithium is only half-filled. This conclusion remains valid if we consider a three-dimensional array of Li atoms.

This picture of metallic bonding readily explains metallic behaviour. Firstly, since the m.o.'s are spread over the entire crystal, electrons in these m.o.'s are completely delocalized. Secondly, the half-filled nature of the electric field, thus, permitting electrical conduction. The same two factors also account for thermal conductivity. The delocalized electrons absorb and re-emit light, rendering the metal surface shiny and lusturous. Finally, since the bonding electrons are not localized at any particular atom, the lattice can be deformed easily, thus, explaining the malleability and ductility of metals.

Our analysis of lithium band structure applies equally well to other alkali metals. If we apply the same analysis to magnesium which has two electrons in 3s level, the 3s-band would be predicted to be completely filled, thus, making magnesium a non-metal. However, the width of energy bands depends upon crystal structure and the crystal structure of magnesium is such that the 3p-band overlaps with the 2s-band. This means that electrical and thermal conduction can take place and magnesium exhibits metallic behaviour.

This example raises the obvious question, "Under what conditions band overlap can be expected?" The answer is also quite obvious. If the orbital interaction (that is orbital overlap) increases, there is greater lowering of energy and also greater overlap of bands involving different orbitals. Orbital irteraction increases with decrease in internuclear distance. One way to shorten the internuclear distance is by applying pressure. Thus, many elements (even non-metallic ones) under high pressure undergo phase transitions and become metallic (conducting).

In fact, calculations show (hat at a pressure of a few hundred thousand atmospheres, the hydrogen atoms should form a cubic close-packed structure which should conduct electricity. Another way that changes in internuclear separation can occur is when the crystal structure of a solid changes.

Thus, grey tin which has relatively open structure (density = 7.29 gm/cc) is a metal. As long as the crystal structure of a metal is not disturbed, its band structure remains unaltered. This is why foreign atoms of the same valency and similar size can readily replace atoms in a metal in arbitrary proportions (forming alloys) without significantly altering the crystal structure or band structure.

BAND MODEL

We shall try to understand it by means of a small ball in a well with steps at different heights. If the ball is rolled on the sides of the well it can either rest at the bottom of well (ground state) or on one of the steps (excited states). If there are more than one balls, the correct situation will be that we start from the bottom up and the number of balls which can be accommodated at each level depends on some parameters that characterize the steps. An atom may be thought of in a well with steps at different heights.

If there are non-interacting atoms, *e.g.*, argon, there would be many such wells separated by interatomic distance whereas in large amount of cases atoms do interact. In that case we have a large number of smaller steps or bands very close to each other between consecutive level. In each band there are as many steps as the atoms. In a diamond gem of one carat there are about 10^{23} atoms, hence, this will be the order of magnitude of steps with different energies. It is not possible to accommodate the atoms outside the well. If the broadening of band is profound, the forbidden zone between two bands disappears and a wider band is obtained.

Let us imagine some small balls in a container. These will fill the box with a lid in flat condition up to a certain level. If we tilt the container in one direction the balls will roll and occupy positions higher up with respect to the bottom of the box. If the box is completely full, tilting does not produce any moment in the balls. If these balls are electrons, at the bottom of the box, the lower energy limit and the lid, the upper energy limit, it can be seen that if the band is not full, the electrons will have a net movement in the direction of the applied field, *i.e.*, they will show electrical conduction.

We can still imagine the balls moving in a full box if they are churned but as many balls will move in one direction as in other. If the band is full, the electrons are not localized in one position, they can move but as many will go in one direction as in other. The net apparent result will be no motion, *i.e.*, no conduction, *e.g.*, diamond.

Since the above model does not depend on the atom, hence, the above model is more universal. The liquid is empty in non-conductors (diamond, neon, etc.), while in conductors (Cu, Na, etc.) partially filled or overlapping of level takes place.

IMPERFECTIONS IN SOLIDS

In crystals, the state of complete order and of lowest energy is found at absolute zero of temperature. At any temperature above 0°K there will be some departure from complete order. In principles, any deviation from a completely ordered arrangement in a crystal constitutes native disorder. In addition to native disorder, crystals may have disorder due to the presence of impurities. The term imperfection or defect is generally employed to denote a departure from a perfectly periodic array of atoms in a crystal, immaterial of whether the imperfection is intrinsic or extrinsic in origin. Many a property of a solid such as the electrical conductivity, optical spectra or mechanical strength cannot be explained on the basis of its crystal structure alone. Imperfections may not only change the properties of a crystal, but may also give rise to certain new properties of interest.

The important types of native disorder are those associated with atomic and electronic imperfections. In addition, orientational disorder and notional disorder are also found in certain solids. A typical example of orientational disorder is found in ferromagnetic solids where all the magnetic moments are alligned in magnetic field at low temperatures. At higher temperature, the solid becomes paramagnetic due to the disorientation of the magnetic moments.

Motional disorder is exemplified by ammonium halides, where the ammonium ions rotate in phase at low temperatures. At higher temperatures the ions rotate out of phase. In this section, we shall be limiting our discussion mainly to the two more important types of native disorder arising from electronic and atomic imperfections.

Electronic Imperfections

It may appear strange at first sight to consider electrons in solids as imperfections, but we will soon see that such a classification is helpful in understanding several phenomena. In perfect covalent or ionic crystals at 0°K, electrons are present in the fully occupied lowest energy states. The distribution of the valence electrons depends on the nature of the chemical bonds in the case of covalent crystals, while they are mainly concentrated about the electronegative component in ionic crystals. In both these cases electrons do not move under an applied electric field. Above 0°K, some of the electrons may occupy higher energy states depending on the temperature and the energy distribution in the allowed states.

Thus, in crystals of pure silicon some electrons are released thermally from the covalent bonds at temperatures above 0°K. These electrons which are free to move in the crystal would be responsible for the electrical conductivity in silicon. The electron deficient bond produced by the removal of an electron is referred to as a positive hole. Holes also give rise to electrical conductivity, but the direction of p motion of the holes in an electric field will be opposite to that of the electrons. Excited electrons and holes in solids are considered to be electronic imperfections. The electrons and holes are generally designated by e and h respectively and their concentrations by n and p.

The concentrations of holes and electrons will be equal in pure crystals like silicon at equilibrium. Electrons and holes can be preferentially produced in covalent crystals like germanium or silicon by adding suitable impurities. Impurities with excess valence electrons (P or As) give rise to extra electrons in germanium and silicon; impurities like Ga or Al with one electron less than Ge or Si produce holes.

In most ionic solids electrons are localized on a particular ion (unlike in metals where electrons are delocalized). The concept of electrons and holes can be employed in ionic crystals just as covalent solids.

This is best understood by considering an ionic crystal where cations of two different valence states can be present. Consider a solid containing mainly ferrous ions.

If we introduce a small number of Fe^{+3} ions in this solid, it is equivalent to introducing holes. This is because, Fe^{+3} ions have one electron less than the preponderant Fe^{+2} ions in the solids. In such a solid, the electrons and holes spend appreciable time at cation sites before 'hopping' to the adjacent sites. The hopping process may be visualized in terms of the change, Fe^{+2} Fe^{+2} Fe^{+3} $\rightarrow$ Fe^{+2} Fe^{+3} Fe^{+2} Fe^{+2}. The hopping process is responsible for the electrical conduction in such solids.

Atomic Imperfections

If imperfections in a crystal are caused by departure from the periodic arrangement in the vicinity of an atom or a group of atoms, the imperfections are called point defects. If the deviations from periodicity extend over microscopic regions of the crystal, they are referred to as lattice imperfections.

Lattice imperfections may extend along lines (line defects) or surfaces (plane defects). Line defects are also called dislocations. The various types of atomic imperfections in crystals are listed in Table 9.

Table 9 : Atomic imperfection in crystals

Imperfection	*Nature of imperfection*
Point defects	
Schottky defect	Atom missing from the normal lattice site creating a vacancy.
Interstitial	Atom in a normally vacant interstitial site.
Anti structure	Atoms misplaced wrongly in site assigned for some other atoms.
Frenkel defect	Atom in the lattice site displaced to an interstitial site creating a vacancy.
Line defects	
Edge dislocation	Dislocation line marks the edge of an extra plane of atoms which is inscrted part way in a crystal.
Screw dislocation	Displacement of atoms in one part of a crystal relative to the rest of the crystal causes a spiral around the dislocation line.
Plane defect	
Grain boundary	Boundary between two crystals in a polycrystalline solid.
Stacking faults	Boundary between two layers with different stacking sequences in close-packed structures.
Shear structures	Structures where excess atoms are present in folded planes connecting slabs of the crystal of normal structure.

Point Defects

When an atom is missing from its normal lattice site a lattice vacancy (Schottky defect) is created. In the case of silver halide, the silver ion is trapped in the interstitial space and thus, creating a hole at some other place.

In the case of alkali halide crystals there are electron traps known as V and F centres. Heating the sodium chloride crystal in vapour of sodium generates F centres, while in the presence of chlorine vapour a υ centre is generated.

An F centre is an electron in a halide ion vacancy and the defect is known as Schottky defect.

In stoichiometric ionic crystals, a vacancy of one ion is accompanied by the creation of a vacancy of the oppositely charged ion in order to maintain etectroneutrality. Thus, equal numbers of cation and anion vacancies are found in alkali halides. In NaCl, there are approximately 10^6 Schottky pairs per cc at room temperature; in one cc there are about 10^{22} ions and therefore, there will be one Schottky defect for 10^{16} ions. Presence of a large number of Schottky defects in a crystal lowers its density markedly. The pyknometric density of vanadium monoxide, VO, is 5.6 g/cc while the x-ray density is 6.5 g/cc. Obviously, there is about 15% disorder of the Schottky kind in VO.

Atoms (or ions) which occupy the normally vacant interstitial sites in a crystal are called interstitials. The important factor determining the formation of interstitials is the size of the atom. There is another kind of point defect called the antistructure disorder, where atoms occupy sites which in the ideal crystal are assigned to atoms of a different kind. In a crystal containing atoms X and Y, some of the X atoms occupy Y sites and an equal number of Y atoms occupy X sites, if there is antistructure disorder in the crystal. Antistructure disorder is commonly found in alloys, particularly at higher temperatures, but is rare in ionic crystals since it would bright like ions next to each other.

In addition to the three basic types of point defects (Schottky, Interstitial and Antistructure) discussed above, three hybrid type of defects can arise from combinations of any two of the basic types as shown below. Of these three combinations the Frenkel defects are more common in ionic solids. In pure alkali halides one does not encounter Frenkel defects since the ions cannot get into the interstitial sites. Frenkel defects are however, found in silver halides where the Ag^+ ions are considerably smaller than the halide ions. Unlike Schottky defects, Frenkel defects do not change the density of the solid. In many real systems various types of point imperfections may occur simultaneously. Thus, in NiAl, Schottky and antistructure disorder are known to be present; in AgBr Schottky and Frenkel defects coexist.

There is a large variety of non-stoichiometric inorganic solids which contain an excess or deficiency of one of the components. Such solids showing deviations from the ideal stoichiometric composition form an important group of defect solids. There are some solids which are difficult to prepare in the stoichiometric composition; thus, the ideal composition in compounds like FeO is metastable (normally we get $FeO_{0.95}O$). The most stable compositions of praseodymium and terbium oxides under ordinary conditions are Pr_6O_{11} and Tb_4O_7 respectively. Similarly, many of the II-VI group compounds such as ZnO and CdS are difficult to prepare in their stoichiometric compositions. In fact, compounds become more interesting to a solid state chemist when there is non-stochiometry. In certain solids, gross variations in composition are also accompanied by changes in crystal structure, while in some others marked deviations from the stoichiometric composition may occur without any change in crystal structure. Thus, VO and TiO can be prepared over a range of compositions (VO_x or TiO_x; x = 0.6 –1.3) with NaCI structure.

Zinc oxide loses oxygen reversibly at high temperatures and turns yellow in colour. The excess metal is accommodated interstitially, giving rise to electrons trapped in the neighbourhood. The enhanced electrical, conductivity of the non-stoichiometric ZnO arises from these electrons (n-type conduction). In FeO, the non-stoichiometry arises from an excess of anions and the electrical conductivity is due to positive holes (p-type conduction). In such oxides the deviation from stoichiometry is accommodated by the change in the valence state of the cation; the cation valency increases if there is anion excess. These two behaviours of oxides can be understood in terms of the basic chemical reactions of the type given below:

Oxygen (anion) excess: $1/2\ O_2(g) = O^{-2}(s)$ + cation vacancy + 2 h (p-type)

Oxygen (anion) deficiency: $O^{-2}(S) = 1/2\ O_2g)$ + anion vacancy + 2e (n-type)

Anion vacancies in alkali halides are readily produced by heating the alkali halide crystals t in an atmosphere of the alkali metal vapour. When the metal atoms deposit on the surface of the alkali halide crystal, halide ions diffuse to the surface and combine with the metal atoms. The electrons produced by the ionization of the alkali metal atoms then diffuse into the crystal and combine with a negative ion vacancy, electrons trapped in anion vacancies (referred to as F centers) give rise to interesting optical and electrical properties.

Thus, excess potassium in KCl makes the crystal appear violet, and excess lithium in LiCI makes it pink. Alkali halides with holes trapped in cation vacancies (V centers) can be produced by heating the alkali halides in an atmosphere of the halogen. Such colour centers can be produced by the irradiation of alkali halides by high energy radiation like x-rays or γ-rays.

Till now the discussion of point defects has been limited to systems where there were no foreign atoms or impurities. Foreign atoms can occupy either interstitial or substitutional sites in a host crystal. The electronic structure of the impurity is of importance in the case of substitutional solid solutions while the size of the impurity atom determines the formation of interstitial solid solutions.

Many of the impurity systems provide new and interesting properties of great importance; the most well-known of such solid solutions are those of Group III or Group V impurities with Group IV elements like Ge or Si. Group III elements like Ca or Al and Group V elements like P or As enter Ge or Si lattice substitutionally. The Group V elements will have one excess valence electron after forming the four covalent bonds normally formed by Group IV elements.

The excess electrons may be released to move through the host structure by thermal energies to give rise to n-type conduction. A Group III element which has only three valence electrons creates one electron deficient bond giving rise to a hole which can move in the crystal carrying an effective positive charge. The electrical conductivity will therefore, be of p-type. In solid solutions of Group IV elements with either Group V or III elements, the overall charge neutrality is maintained. The impurities in these cases are ionized: imperfections that ionize releasing electrons (*e.g.*, P or As in Ge or Si), are called donors while those which ionize by accepting electrons (*e.g.*, Ga or Al in Ge or Si) are called acceptors.

The magnitude of energy required to ionize impurity atoms is generally small and in germanium at ordinary temperatures (0.025 eV) nearly all the donor and acceptor levels are ionized. The ionization energy decreases with the increase in impurity concentration and if the concentration is sufficiently high, a large fraction of ionized donors and acceptors is found even at low temperatures. The extra electron of the donor atom may be visualized as moving in orbits in the field of a positive charge.

The picture is somewhat similar to that of the hydrogen atom, the main difference being that the electron and the positive charge are in a medium of high dielectric constant. Further, the radius of the orbit is much larger covering several atomic distances.

A common method of introducing defects in ionic solids is by adding impurity ions. If the impurity ions are in different valence state from that of the host ions, vacancies are created. Literature abounds in examples where interstitials or vacancies are produced by addition of impurities to ionic solids. For example, addition of $CdCl_2$ to AgCl or $SrCl_2$ to Nad yields solid solutions where the divalent cations occupy the Ag^+ or Na^+ sites and produce cation vacancies equal in number to that of the divalent ions. If one adds a molecule of a rare earth fluoride, MF_3 to CaF_2, the type of the concentration of defects produced will depend on the valence state of the rare earth ion in the host lattice. If the rare earth ion is in the +3 state, one F^- interstitial will be formed beside a new cation site; if the rare earth ion is in the +2 state there will be no interstitial F^- ion. The valence state and the defect structure in such systems is best studied by spectroscopic methods.

If the ions in a host crystal can assume two or more valence states, addition of an impurity ion in a valence state different from the normal valence state of the ions in the host crystal produces changes in the valence state of a controlled number of host ions. Such controlled valency systems show interesting semiconducting properties. A good example of a controlled valency semiconductor is NiO doped with Li_2O where the number of N^{+3} ions in the system is determined by the number of Li^+ ions. The composition of the resulting material may be written as $Ni^{+2}_{1-2y}\ Ni^{+3}_{y}\ Li^{+}_{y}O$. With 10% atoms of Li. This material is black in colour and exhibits high electrical conductivity (p-type); NiO is pale green in colour and is an insulator.

The most important criterion in the defect chemistry of ionic crystals is the principle of electroneutrality or charge compensation. The creation of vacancies or new valence states of ions in crystals by the addition of impurities is determined by this simple principle.

SEMICONDUCTORS

Solids, containing ions or ion vacancies which move under the influence of applied field are known as semiconductors. Their conductivity is intermediate between metals and insulators and is of the order 10^2 to 10^{-10} mho cm^{-1}. The electrical conductivity in metals decreases with the increase in temperature but in semi-conductors it increases with the increase in temperature. Their conductivity is either intrinsic or due to impurities.

Unlike metals, the conductivity of semiconductors and insulators are mainly determined by the impurities and defects. Electrons and holes produced by the ionization of defects contribute to the electronic conduction in these solids. The behaviour of insulators and semi-conductors can be explained in terms of the band model. If the band structure of an element is such that the fully occupied band is separated by a large energy gap from the vacant band, then conduction is not possible. This situation is found in good insulators like diamond or MnO. On the other hand, if the energy gap is small, promotion of electrons can take place by raising the temperature; higher the temperature greater is the number of electrons promoted and greater is the conductivity. This is characteristic behaviour of semi-conductors (*e.g.*, SnO_2; $SrTiO_3$).

In Table 4.9 we have summarized the electrical properties of some oxides of transition metals to illustrate the wide variations found in such materials. It is particularly interesting that the monoxides all of which possess the NaCI structure show such marked differences in electrical properties; this is obviously related to electronic configurations of the transition metal ions.

In addition to electronic semiconductors where the band picture is applicable, there are many materials where the semiconduction is due to the hopping of electrons from one site to another. Examples of these are oxides where cations of more than one valence state are present as in Pr_6On, NiO doped with LiO, etc.

Table 10 : Electrical properties of typical transition metal oxides

TiO(M)	VO(M)	CrO(?)	MnO(I)	FeO(s)	CoO(s)	NiO(s)	CuO(s)
Ti_2O_3(M-S)	V_2O_3(M-S)	Cr_2O_3(s)	Mn_2O_3(s)	Fe_2O_3(s)	CO_3O_4(S)	–	Cu_2O(s)
TiO_2(s)	VO_2(M-S)	CrO_2(M)	MnO_2(D)	Fe_3O_4(D)			
	V_2O_5(S)		Mn_3O_4(s)				

M = Metal; I = Insulator; S = Semiconductor; M-S = Shows a transition from metal to semiconductor behaviour at a certain temperature; ? = not known; D = Degenerate semiconductor (high conductivity with little or no variation with temperature).

Ionic conduction in a solid involves migration of atoms, ions or other charged species under applied field and it is generally useful to study the diffusion (migration due to a concentration gradient) of these species in the solid. Conductivity measurements in ionic solids like NaCI give valuable information on the energy required for the creation of defects, as well as their migration energies.

INSULATOR

The number of electrons promoted will increase exponentially with temperature. Thus, an insulator can in principle be made more conductive by thermal excitation of electrons but if the energy gap is so large that the number of electrons excited even at high temperature is negligible, the material is classified as insulator. If the energy gap is very small there will be an effective thermal excitation at low temperatures.

THERMAL EFFECT

The effective average energy of the electron in thermal equilibrium with the surroundings is called Fermi energy. In the case of metals it is the energy of the electrons that can be thermally promoted to a conduction state. At absolute zero it is the energy of the most energetic electron. In the case of insulators it lies about the topmost valence band.

The thermal excitation of the electrons in an insulator provides a few conduction electrons in the upper band and in the lower band an equal number of mobile holes are generated. The energy of these current carrying holes and electrons lies in the energy gap that separates them.

When the thermally produced electrons and holes are mainly responsible for the conductivity of a semiconductor (generally at higher temperatures) rather than the electrons and holes produced by donors and acceptors (as in doped Ge or Si), the region is called the intrinsic region. At lower temperatures where the conductivity is mainly determined by the concentration of donors and acceptors, the region is called extrinsic. We should recapitulate here that pure germanium or silicon by itself would be an insulator; only by doping with group III or V impurity can we increase the conductivity at ordinary temperatures.

The intrinsic conductivity in a semiconductor is due to the thermal excitation of electrons from valence band into empty states of the conduction band with energy gap of the order of RT or 600 cal mole^{-1}. As this energy gap approaches zero, valence and conduction band merge and the conduction becomes metallic Fig. 2.34a(ii).

Impurity conductivity is either due to foreign atoms or lattice imperfections.

If pure Si or Ge is exposed to the vapours of P or As, they are held into the lattice of Si or Ge. The extra electron associated with each foreign atom remains near by and seeks a state of low energy in the range of energies forbidden to electrons in pure Si or Ge. Since they carry negative charge they are called n-type of semiconductor. In the case of Si, the minimum energy required to transfer an electron from an impurity centre to conduction band is 1.84 Kcal J mole^{-1}.

If B, A*l*, or Ga are held into the lattice of Si or Ge each foreign atom will be associated by electron valencies or holes. These electrons are mobile. When a hole is filled by an electron, a hole is created in some other position. So in effect these holes move contributing to the conductivity of the solid. Since they are positive charge carriers, they are called p-type of semiconductors. These semiconductors are used as solid state electronic component called transistor.

ELECTRICAL PROPERTIES

Based on their electrical conductivity, solids can be broadly classified into three types: metals, insulators and semiconductors. Conductivity of solids varies anywhere from 10^8 ohm^{-1} cm^{-1} in metals to 10^{-22} ohm^{-1} cm^{-1} in insulators. Electrical conductivity of solids may arise through the motion of electrons and holes (electronic conductivity) or other charged imperfections and ions (ionic conductivity). Substances like pure alkali halides, where the conduction is through ions only are generally insulators since the mobilities of ions are very much lower than those of holes or electrons. However, presence of vacancies or other imperfections could markedly increase the conductivity of ionic solids.

At laboratory temperatures, conductivity of metals is nearly independent of impurities and lattice defects. The electron concentration is mainly determined by the nature of the metal (number of valence electrons in the metal) and the mobility by the lattice vibrations. At low temperatures, however, the lattice vibrations are not limiting and the conductivity should be infinite. However, this is not true because of the presence of lattice imperfections and impurities. It is thus, convenient to use the resistance ratio p 300°K/ρ. 4.2°K as a measure of purity of the metal.

LINE DEFECTS

Although the concept of dislocations was first introduced primarily to explain deformation in metals, it is now established that dislocations play an important role in crystal growth, crystal transformations, electrical conductivity and other solid state phenomena. Dislocations in crystals have been observed directly with the aid of electron microscopes.

Edge dislocations where an extra half-plane of atoms is inserted into the lattice is illustrated. The dislocation consists of a line of atoms each of which is short of one coordinating atom than prescribed by the crystal structure. The dislocation extends along a direction perpendicular to the plane of the paper and is designated by the symbol ⊥. The dislocation line marks the edge of the plane of atoms inserted part-way into the crystal. Edge dislocations are also called Taylor dislocations or Taylor-Orowan dislocations.

Another kind of dislocation called the screw dislocation described by Burgers (also known as Burgers dislocation) consist of a line of atoms with distorted coordination polyhedra. This is caused by a displacement of atoms in one part of the crystal with respect to the rest of the crystal, the displacement direction being parallel to the dislocation line. The row of atoms marks the termination of the displacement. The polyhedron is not regular in the disturbed region and the distortic decreases progressively with increasing distance from the dislocation line.

(a) *Structure of an edge dislocation :* The deformation Is the result of inserting an extra plane of atoms on the upper half of the V-axis. Atoms in the upper half-crystal are compressed and those in the lower half are extended.

(b) Model to show termination of an edge dislocation in Sb of the surface of the crystal (H.C.Gatos, M.I.T.)

(c) Electron microscope picture of a thin crystal of platinum phytholocyanine (direct transmission) showing crystal planes and the presence of an edge dislocation. Magnification (300.000). (J.W. Menter, Tube Investments Research Laboratories).

An edge dislocation can be produced by applying a shear to the crystal and the net displacement of atoms in one part of the crystal relative to the atoms in the other part of the crystal determines the magnitude of the dislocation. Since atoms in acrystal occupy equilibrium positions, the net displacement is always equal to one of the equilibrium distances between atoms. If this distance (Burgers vector) is equal to the unit lattice translation, the dislocation is referred to as a unit dislocation. If the Burger's vector is shorter than a unit translation, the dislocation is called a partial dislocation.

PLANE DEFECTS

Many of the solid state materials are polycrystalline in nature. Polycrystalline solids consists of a number of small randomly-oriented interlocking crystals (grains). The grain boundary generally has a complex structure which in some way conforms to the shapes and orientations of the adjacent grains. Since a grain boundary forms a discontinuity in the periodicity of the lattice, it may be considered to be another kind of imperfection.

The disturbed region in the boundary layer may consist of a few atom layers and acts as a bridge between the adjacent grains. These disturbances are generally large due to random orientation of the initial grains. Since grain boundaries may contain a number of imperfections like impurities, interstitials, etc. (expelled by the grains during their growth), it often becomes difficult to understand the electrical, mechanical and other properties of polycrystalline solids.

Another important kind of plane defect found in crystals arises from stacking faults. In the cubic close-packed structures, we have the sequence ... ABCABC ..., while in the hexagonal case we have the sequence ... ABAB ... A fault in a stacking sequence as in ABABBCBCBC or ABCABCBCABCABC, occurs when the packing is cubic instead of hexagonal or vice versa. The sequence in the planes before and after the stacking fault would, however, be normal.

4

Statistical Mechanics

INTRODUCTION

The word particle is used in a broad sense meaning a fundamental particle *e.g.*, an electron, an atom, a molecule etc. The particle is a well defined and a stable unit of a given physical system. As it is not practically possible to determine the property of each particle individually, a statistical approach is made by using the concept of probability of distribution. This statistical approach helps in determining the bulk or macroscopic property of the system as a whole. The idea of probability does not imply that the particles move in a random way without obeying the laws. For a many particle system, the statistical analysis is valid regarding the applications of probability distribution of particles. Statistical mechanics deals with systems consisting of many particles and the methods employed are to get a collective or macroscopic property of the system without taking into account the individual motion of the particles. Statistical mechanics acts as a bridge between thermodynamics and quantum mechanics. The classical thermodynamics is based on empirical laws in which the variables such as pressure, temperature and energy are regarded as functions of state. However, in the derivation of relations of thermodynamic functions, we ignore the constitution of the material system. But the experimental observations support the laws.

The only drawback in classical thermodynamics is that we cannot calculate directly the absolute value off any thermodynamics function of a substance. In order to over come this, a new method of approach has been developed and is known as statistical thermodynamics. The statement according, to which non-spontaneous process also called negative process cannot be the only result of a complex of processes is not strict are negative processes in microscopic (bulk) system events. Therefore the second law of thermodynamics first law, but a statistical law that is observed with a high degree off accuracy for considerable numbers of molecules and is less applicable, the smaller are the dimensions of the system being studied. So the second law of thermodynamics has statistical nature. Classical thermodynamics has bee applied successfully to relate the microscopic properties of the individual molecules (*e.g.* moment of inertia, dipole moment etc.) with macroscopic properties (*e.g.* molar heat capacity, polarization, etc.) of a system having a large number of molecules.

In supposed to be composed of innumerable particles in motion. This motion of particles is governed by the laws of dynamics and in fact subject to quantum the number of particles is so large that it is impossible to follow the behaviour of every individual particle Statistical methods can be used to find out the properties of the system, particularly to know what fraction of the molecules has a given energy.

It is possible to establish that the laws of thermodynamics are the necessary consequence of the quantum principles, when statistical postulates are used. In this method, we take into account the atomic constitution of matter. The matter is supposed to consist of innumerable particles in motion is governed by the laws of dynamics. Quantum mechanics provides information about the energy of the molecular system, statistical mechanics tells us about the possible arrangement of the energy among various molecules of the system and introduces the concept of probability and partition functions. Statistical thermodynamics deals, with the relationships between the probability partition functions and thermodynamic properties. In other words, a physical entity generally manifests itself in :

(i) A Microscopic, and

(ii) A Macroscopic state.

The dealt with macroscopic or the bulk state of matter in the preceding chapters.

The macroscopic variables were pressure, volume, temperature, entropy, etc. The macroscopic variables were related mutually on thermodynamical considerations. Further, we assumed that the number of atoms/molecules in them (microscopic details of a macrosystem) were exorbitantly large $\approx 10^{25}$ m^{-3}. The microscopic systems, on the other hand, have very few particles. Now, we know that the actual systems, the macro systems, are formed of the microsystems and so there should be a definite relation between the two.

This fact was suggested by L. Bottzmann. He explained in this way the contradiction between the reversibility of mechanical motion and the irreversibility and unidirectional nature of real physical and chemical processes; this unidirectional nature is the result of the molecular structure of the material world. In this way, it was Boltzmann who provided the molecular basis to chemical thermodynamics. This approach is called chemical statistics or statistical thermodynamics and it deals with the applications of mechanics to chemical thermodynamics. If we want to have a complete description of a system we need to know actual velocity (or momentum), location and mutual interaction (if any, with other particles) of every particle. Since every panicle is described by three positions and three momentum coordinates, hence to describe a macrosystem we need a very large number of data. Further, any change in the parameter of an individual particle should be reflected in the properties of the system as a whole. *This poses a question*: How to deal with such a vast data which even a super-computer is not capable to handle, if at all we are able to collect it. If, however, we remember that:

(i) Much of the data and analytical results obtained on micro level are not useful in the study of the macrosystem, and

(ii) The application of classical dynamical principles to a many particle system is theoretically inapplicable, we can easily dispense with the technically impossible and practically useless classical kinetic and thermodynamic methods in favour of new statistical methods.

In the following sections we will see the importance of the concept of probability in the development of the statistical theory. Hence, we shall first consider some elementary theorems on the calculus of probability. In other words, it refers to a class of events of which the particular event is one member. Mathematically, the probability is defined as: When a certain aspect is possible in a ways and is not possible in b ways, then the probability of occurring the aspect is a/(a + b). (Newton defined probability as the ratio of a certain number of aspects, which were possibly observable, to the total number of aspects included in that class.) Here (a + b) is the total number of aspects each of which has the same chance of occurrence.

Statistics is applied only to such systems which contain a large number of identical individuals which may be distinguishable by means of some property. In a statistical method, we study the behaviour of aggregates of particles to calculate the average value without investigating the detailed behaviour of the system and finding there from the value for each individual. Statistically, it is sufficient to know the most probable distribution of particles in different accessible states. This branch of physics has seen its development in the hands of Maxwell, Boltzmann, Gibbs, Fermi-Dirac, Bose-Einstein etc.

Now, as we have shown, if an event can occur in a ways, the probability of occurring of this event is a/(a + b), and similarly, if it fails to occur in b ways, then the probability of its failing is b/(a + b). The sum of the two

$$\frac{a}{a+b} + \frac{b}{a+b} = 1$$

corresponds to the certainty of some event, since the event must occur or fail to occur. If b = 0, the probability of occurrence becomes certain but if a = 0, the probability of its failure is certain.

We may define the probability for a certain event occurring m times in n unrestricted (free) attempts as,

$$P = \lim_{n \to \infty} \frac{m}{n}.$$

MACROSCOPIC SYSTEMS CHARACTERISTIC FEATURES

Let us now discuss some of the characteristic features of the macrosystems. In doing so we will assume that it has a large number of microscopic particles.

Fluctuation

Let the isolated system has N molecules of a perfect gas and that it was left to itself for quite some time so that it is in equilibrium. Further, also consider that the system has been partitioned into two identical (also equal) parts by an imaginary wall, each part A and B has n and n molecules respectively such that n + n = N.

Now, as we have seen, for large values of N, $n \cong n^1 = N/2$

But the molecules of the gas are in a state of constant random motion. So that some of the molecules are leaving portion .A and moving into B and vice-versa. Thus, there is a continuous change in the number of molecules in each portion of the box. Time variation of this change is known as fluctuation. The state of normal fluctuation of an isolated system. Usually, N/2 – n is only a small number. For large deviation, the number of microstates (and hence the probability) reduce drastically. The number of microstates for different values of N and N/2 – n. For the sake of simplicity the deviation N/2– n is represented as a fraction of total number of particles. Value of R ($\cong P_x/P_{max}$) is also shown to arrive at qualitative results;

In actual physical problems the number of molecules are $\sim 10^{24} m^{-3}$ whence even 0.01% deviation also becomes quite improbable. This is the reason for quick diffusion and most equitable distribution of air molecules in the available space.

STATISTICS APPLIED TO MOLECULES

A monatomic perfect gas. The state of each molecule of the gas in the μ-space is represented by a point (and by a cell in quantum mechanical terms). But, since the gas molecules are in constant motion, their state will be changing continuously in space and, due to collisions (and also mutual interactions), there is a change .in momentum state also. The trajectory of such a molecule constant energy surface given by

$$E = \frac{p_x^2 + p_y^2 + p_z^2}{2m} = V(q_x, q_y, q_z)$$

The real gas molecules, however, are endowed with mutual interactions in addition to usual collisions, hence their state is better represented by a point in y-space. The rate of change of its state, then, depends on the quantum of interactions. The path of representative point in y-space is given by the relation

$$E = \sum_i T_i + \sum_i V_i + \sum_{i,j}{}' U_{ij}$$

where T_i, V_i and U_{ij} refer to kinetic energy, potential energy and energy of interaction of ith particle. The prime on the summation sign suggests that i = j is not allowed (which would mean self-interaction). With this background and remembering that the number of molecules is very large in a macrosystem, we now proceed to investigate the distribution details.

SOME ENSEMBLES

It would be appropriate to classify the ensembles according of their dependency on macroparameters. Since an ensemble is an assembly of macroscopically identical but mutually independent systems, a change in macroparameter (like volume, pressure, energy, number of molecules, etc.,) may lead to another set of systems:

Canonical Ensemble

In this ensemble all its members are identical and remain in thermal equilibrium with the source (*i.e.*, members of the ensemble can exchange thermal energy with the reservoir). Thus all the members of a canonical ensemble have the same number of particles N, volume V and are at the same temperature T. A canonical ensemble. Its members are shown separated by impermeable but conducting rigid, walls and they are in thermal contact with the source. Gibbs gave a relation,

$$\rho(\varepsilon) = Ae^{-\varepsilon\tau}$$

for density of its phase points. A and T in the relation are constants. The ensemble can be viewed physically as a freezer containing water filled metal-ice cube tray.

Grand Canonical Ensemble

The members of this ensemble can interact thermally, mechanically and also diffusively with each other and with the reservoir. Hence they are at the same temperature of same volume and at same chemical potential such an ensemble. Each member here is separated by flexible, permeable and conducting walls, which help exchange mass and energy between its members and with the surroundings. The constituent molecules in the member systems are thus of different kinds (different chemical species). The density of phase points in such ensemble is

$$\rho(N) = e^{(\Omega+N\mu-\varepsilon)/\tau}$$

where, Ω, μ and τ are constants. A grain of sugar in boiling water can best be viewed as a grand canonical ensemble.

Uniform Ensemble

Those ensembles which have same density in phase space, are called *uniform ensembles*.

Hence, if ΔN represents the number of system, in volume element $d\Gamma$ we have, for a uniform ensemble,

$$\underset{d\Gamma\to 0}{\text{limit}} \frac{1}{d\Gamma}\frac{\Delta N}{N} = \text{constant}$$

Microcanonical Ensemble

In these ensembles the energy of their member systems remain constant, *i.e.*, they do not interact with each other or with the surrounding. Thus, they are free from external effects. The ensemble is thus conceptually an idealised one and can be represented by a function that depends only on the energy. An ensemble, here ε and V are constant. Every member is separated from the other by rigid, impermeable and insulated walls. The density of phase points within each member system is uniform and is zero at the walls. Physically, we may view the microcanonical ensemble as a perfect insulating enclosure encompassing several isolated systems. It is suitable for ideal gases.

SOLAR CONSTANT

The source of heat (sun) radiations and it emits heat radiations in all directions. The earth receives only a fraction of the energy emitted by the sun. The atmosphere also absorbs a part of the heat radiations and air, clouds, dust particles etc., in the atmosphere scatter the heat and light radiations falling on them. From the quantity of heat radiations received by the earth, it is possible to estimate the temperature of the sun. Therefore, to determine the value of a constant, called solar constant, certain ideal conditions are taken into consideration. *It is the amount of heat energy (radiation) absorbed per minute by one sq cm of a perfectly black body surface placed at a mean distance of the earth from the sun, in the absence of the atmosphere, the surface being held perpendicular to the sun's rays.*

The instruments used to measure the solar constant are called *pyrheliometers*. The heat energy absorbed by a known area in a fixed time is found with the help of the pyrheliometers. To eliminate the effects of absorption by the atmosphere, the value of the solar constant is found at various altitudes of the sun on the same day under similar sky conditions. If S is the observed solar constant, S_0 the true solar constant and Z the altitude (angular elevation) of the sun, then

$$S = S_0 \, a^{\sec} \, z \qquad \text{...(i)}$$

$$\log S = \log S_0 + \sec Z \log a \qquad \text{...(ii)}$$

Here a is a constant.

A graph is plotted between log S along the y-axis and sec Z along the x-axis. The graph is a straight line. Produce the graph to meet the y-axis. The intercept on the y-axis gives log S_0.

From the value of log S_0, the value of S_0 the solar constant can be calculated. The value obtained varies between 1.90 and 2.60 calories per sq cm per minute.

Temperature of the Sun

The sun consists of a central hot pardon surrounded by the photosphere. The central portion has a temperature of the order of 10^7 K. The photosphere has a temperature of about 6000 K. This temperature is also called the effective temperature of the sun. Considering the sun as a perfect black body radiator, the temperature of the sun can be calculated.

Let the mean distance of the sun from the earth be R and S the solar constant. Then, the total amount' of heat energy received by the sphere of radius R in one minute = $4\pi R^2S$.

If r is the radius of the sun, then the amount of heat energy radiated by 1 sq cm surface of the sun in one minute

$$E = \frac{4\pi R^2 S}{4\pi r^2} = \left(\frac{R}{r}\right)^2 \times S$$

Taking $R = 148.48 \times 10^7$ km

$$r = 6.928 \times 10^6 \text{ km}$$

The mean value of

$$S = 1.94 \text{ cals per cm}^2 \text{ per minute.}$$

$$E = \left(\frac{148.48\times10^7}{6.928\times10^5}\right) \times \frac{1.94}{60} \text{ cal per second} \quad ...(i)$$

Also $E = \sigma T^4$

But $\sigma = 5.75 \times 10^{-5}$ ergs per cm^2 per second

$$= \frac{5.75\times10^{-5}}{4.2\times10^7} \text{ cal per cm}^2 \text{ per second}$$

$$E = \frac{5.75\times10^{-5}}{4.2\times10^7}.T^4 \quad ...(ii)$$

Equating (i) and (ii)

$$\left(\frac{5.75\times10^{-5}}{4.2\times10^7}\right)T^4 = \left(\frac{148.48\times10^7}{6.928\times10^5}\right) \times \frac{1.94}{60}$$

$$T = 5730 \text{ K.} \quad ...(iii)$$

This temperature gives the effective temperature of the sun acting as a black body radiator. The actual temperature of the sun is higher than this value. The temperature of the sun is usually taken as 6000 K.

Temperature of the sun can also be calculated from Wien's displacement law,

$$\lambda_{max} T = 0.2892$$

The wavelength of the radiations for which the energy is maximum in the spectrum is 4900×10^{-8} cm.

Substituting the value of λ_m, the value of T comes out to be 5902 K. This value is in agreement with the accepted value. Hence, the effective temperature of the sun (photosphere) is about 6000 K.

STEFAN'S LAW (DERIVATION OF MATHEMATICAL)

The fact that blade body radiations exert pressure similar to a gas, helps in applying thermodynamics to heat radiations.

Let ψ be the energy density of radiations inside a uniform temperature enclosure at temperature T. P is the pressure and V is the volume.

Applying the first law of thermodynamics

$$\delta H = dU + P.dV \qquad \text{...(i)}$$

Applying thermodynamical relation

$$\left(\frac{\partial H}{\partial V}\right)_r = \left(\frac{\partial P}{\partial T}\right)_v \qquad \text{...(ii)}$$

$$\left(\frac{\partial U + P\partial V}{\partial V}\right)_r = T\left(\frac{\partial P}{\partial T}\right)_v$$

$$\left(\frac{\partial U}{\partial V}\right)_r = T\left(\frac{\partial P}{\partial T}\right)_v = P \qquad \text{...(iii)}$$

Now $\quad U = V\psi$

and $\quad P = \frac{\psi}{3}$

or $\quad \left[\frac{\partial U}{\partial V}\right]_r = \psi$

Here ψ is a function of temperature alone.

Substituting these values in equation (iii),

$$\psi = \frac{T}{3}\frac{d\psi}{dT} - \frac{\psi}{3}$$

$$\frac{4\psi}{3} = \frac{T}{3}\frac{d\psi}{dT}$$

or $$\frac{d\psi}{\psi} = 4\frac{dT}{T}$$

Integrating, $\log \psi = 4 \log T + \text{constant}$

or $$\psi = KT^4 \qquad ...(iv)$$

Here K is a constant

Also the total rate of emission per unit area of a black body is proportional to the energy density.

$\therefore$ $$R \propto \psi \propto T^4$$

$\therefore$ $$R = \sigma T^4 \qquad ...(v)$$

where σ is Stefan's constant.

The value of Stefan's constant in C.G.S. system is 5.672×10^{-5} C.G.S. units and in M.K.S. system it is 5.672×10^{-8} M.K.S. units.

COOLING FROM STEFAN'S LAW (DERIVATION OF NEWTON'S LAW)

Stefan's law is applicable for all temperatures of a hot body. But Newton's Law is applicable when the difference of temperature between the hot body and the surrounding is small. Consider a hot body at a temperature T_1 placed in a uniform temperature enclosure at T_1. According to Stefan's law,

$$R = e\sigma(T_1^4 - T_1^4) \qquad ...(i)$$

Here e is the emissivity of the surface of the hot body

$$R = e\sigma(T_1 - T_2)(T_1^3 + T_1^2 T_2 + T_1 T_2^2 + T_2^3)$$

As $(T_1 - T_2)$ is small, T_1 can be taken approximately equal to T_2.

Then, $$R = e\sigma(T_1 - T_2)(T_2^3 + T_2^3 + T_2^3 + T_2^3)$$

$$R = 4e\sigma T_2^3((T_1 - T_2)$$

Taking $$4e\sigma T_2^3 = k$$

$$R = k(T_1 - T_2)$$

or $$R \propto (T_1 - T_2). \qquad ...(ii)$$

This equation represents Newton's law of cooling and is true when the difference of temperature is small.

PHOTON GAS

The interaction of electromagnetic radiations with matter, led to the idea that electromagnetic radiations are composed of discrete energy particles called photons. Each photon has an energy $h\nu$ and momentum h/λ. Here ν is the frequency and λ is the wavelength of the radiations. The electromagnetic radiations trapped in a cavity and in thermal equilibrium with the walls of the cavity are termed as black body radiations. In the equilibrium condition, the black body radiations can be considered as the *photon gas.* It is assumed that the photons do not interact among themselves. The photons interact only with-the atoms of the walls of the cavity. It is further assumed that the photons are indistinguishable and many photons can have the same energy. Photons are taken as bosons and they obey Bose-Einstein statistics.

As the photons can either be emitted or absorbed by the atoms of the walls of the cavity, the number of photons is not constant *i.e.*, the conition $\sum_i dn_i = 0$ is no longer valid.

Due to this reason the value of α is equal to zero in the Bose-Einstein distribution law.

$$n_i = \frac{g_i}{[e^{(\alpha + E_i/kT)}] - 1} \quad ...(i)$$

Here $\alpha = 0$

$$\therefore \quad n_i = \frac{g_i}{[e^{E_i/kT}] - 1} \quad ...(ii)$$

In case, the cavity is large as compared to the wavelength of the radiations, the energy spectrum of the photons is taken to be continuous. In this case, the energy difference between successive allowed energy value is very small. Thus, replacing g_i by $g(E)\, dE$, and n_i by dn

$$dn = \frac{g(E)dE}{\left(e^{E/kT)}\right) - 1} \quad ...(iii)$$

As the energy of a photon, $E = h\nu$, the value of $g(E)\, dE$ can be taken equal to $g\,(\nu)\, d\nu$. The factor $g(\nu)\, d\nu$ corresponds to the number of oscillatory modes in the frequency range $d\nu$ and relating to the energy range dE.

The number of states in a black body radiation in the frequency range ν and $\nu + d\nu$ can be obtained by calculating the spherical volume bound by the spheres of radii $\frac{h(\nu + d\nu)}{c}$ and $\frac{h\nu}{c}$

The volume of the spherical shell

$$= \frac{4}{3}\pi\frac{h^3}{c^3}(\nu + d\nu)^3 - \frac{4}{3}\pi\frac{h^3}{c^3}\nu^3$$

$$= \frac{4}{3}\pi\frac{h^3}{c^3}[\nu^3 + 3\nu^2 d\nu + \ldots - \nu^3]$$

$$= \frac{4}{3}\pi\frac{h^3}{c^3} \times 3\nu^2 d\nu$$

$$= 4\pi \frac{h^3}{c^3} \nu^2 d\nu \qquad \text{...(iv)}$$

The phase space has volume $V = h^3$ and there are two states of polarization for the radiation. The number of states in the black body radiation in the frequency range ν and $\nu + d\nu$ is given by

$$g(\nu)d\nu = \frac{8\pi\nu}{c^3}\nu^2 d\nu \qquad \text{...(v)}$$

$$dn = \left(\frac{8\pi V\nu^2 d\nu}{c^3}\right)\frac{1}{\left(e^{h\nu/kT}\right) - 1} \qquad \text{...(vi)}$$

For dn photons in the frequency range $\nu + d\nu$ and ν, the energy is equal to $(h\nu)$ dn and the energy per unit volume

$$= \frac{(h\nu)dn}{\nu}$$

Therefore, the energy density distribution for black body radiation is given by

$$E(\nu) = \frac{(h\nu)dn}{V.d\nu} \qquad \text{...(vii)}$$

Substituting the value of dn from equation (vi), we get

$$E(\nu) = \frac{8\pi h\nu^2 d\nu . h\nu}{c^3 V.d\nu\left[\left(e^{h\nu/kT}\right) - 1\right]}$$

$$E(\nu) = \left(\frac{8\pi h\nu^3}{c^3}\right)\left[\frac{1}{\left(e^{h\nu/kT}\right) - 1}\right]^{d\nu} \qquad \text{...(viii)}$$

Equation (viii) represents Planck's radiation law for black body radiations.

From equation (viii)

$$E(\nu)d\nu = \frac{8\pi h\nu^3}{c^3}\left[\frac{1}{\left(e^{h\nu/kE}\right)-1}\right]^{d\nu}$$

But $$\nu = \frac{c}{\lambda}$$

or $$d\nu = -\frac{c}{\lambda^2}d\lambda$$

Neglecting the negative sign, we get

$$\therefore \qquad E(\lambda)d\lambda = \frac{8\pi h\nu^3}{c^3} \times \frac{c}{\lambda^2}\left[\frac{1}{\left(e^{h\nu/kT}\right)-1}\right]^{d\lambda}$$

or $$E(\lambda) = \frac{8\pi hc}{\lambda^5} \times \frac{c}{\lambda^2}\left[\frac{1}{\left(e^{h\nu/kT}\right)-1}\right] \qquad \text{...(ix)}$$

This gives the energy density for wavelength λ in the spectrum of the black body. Both the equations (viii) and (ix) represent Planck's radiation law for black body radiation.

GIBBS FUNCTION

The Gibbs function G of a system is given by,

$$G = U - TS + PV$$

Consider a system that can do other forms of work, in addition to P.dV work, *e.g.*, a voltaric cell. In the case of a voltaic cell, the electrical work is –E.dI. Similarly for a magnetic material, the magnetic work is –m.dH. In general, the work will be given by P.dV plus a sum of terms each being the product of intensive variable (such as P, E or m) and the differential of an extensive variable (such as dV, dI and dH). In the case of a voltaic cell, the intensive variable is S and differential of extensive variable is dI.

Suppose, in general, in addition to PdV the intensive variable is y and differential of extensive variable is dx. The work done for any reversible process,

$$\delta W = PdV + ydx$$

$$W = \int_{V_1}^{V_2} PdV + \int_{x_1}^{x_2} ydx$$

Take $\int_{x_1}^{x_2} y dx = A$

Consider a process where the system works at constant pressure P_0 and the change in volume is $(V_2 - V_1)$

$$\therefore \quad \int_{V_1}^{V_2} PdV = P_0 [V_2 - V_1]$$

$$\therefore \quad W = P_0[V_2 - V_1] + A \quad \text{...(i)}$$

For a system that exchanges heat with a reservoir temperature

$$W < (U_1 - U_2) - T_0 (S_1 - S_2) \quad \text{...(ii)}$$

$$P_0[V_2 - V_1] + A < (U_1 - U_2) - T_0 (S_1 - S_2)$$

or $A < (U_1 - U_2) - T_0 (S_1 - S_2) + P_0[V_1 - V_2]$ ...(iii)

Consider a specific process, where the initial and the final states of the system and the surroundings are at the same temperature (T_0) and pressure (P_0)

$$To = T \text{ and } P_0 = P$$

From equation (iii)

$$A_{P,T} < (U_1 - U_2)_{P,T} - T(S_1 - S_2)_{P,T} + P[V_1 - V_2]_{P,T} \quad \text{...(iv)}$$

But, for the Gibbs function,

$$G = U - TS + PV \quad \text{...(v)}$$

Therefore, for two equilibrium states at the same pressure and temperature.

$$[G_1 - G_2)_{P,T} = (U_1 - U_2)_{P,T} - T[S_1 - S_2]_{P,T} + P[V_1 - V_2]_{P,T} \quad \text{...(vi)}$$

From equations (iv) and (vi)

$$A_{P,T} < (G_1 - G_2)_{P,T} \quad \text{...(vii)}$$

Thus, the difference between Gribbs function of a system between two equilibrium states sets the maximum limit to the work in addition to PdV work, provided the initial and the final states are at the same pressure and temperature and the system exchanges heat with a single heat reservoir. The work done will be maximum when the process is reversible. The process in this case will be isothermal–isobaric. If the process is irreversible, work done will be less than the maximum.

CRITICAL CONSTANTS

(i) ***Critical temperature of a gas*** (T_c) *may be defined as the minimum* *temperature which must be reached before it can be liquefied by the application of pressure.* A gas cannot be liquefied by pressure above its critical temperature, no matter how high the applied pressure may be. Each gas is having definite critical temperature, *e.g.*, critical temperature of CO_2 is 31.1MC.

(ii) ***Critical pressure of a gas*** **(P_c)** *may be defined as the minimum pressure necessary for the liquefaction of a gas at critical temperature.*

(iii) ***Critical volume of a gas*** **(V_c)** *may be defined as the volume occupied b 1 mole of the gas or liquid under critical pressure and under critical temperature.*

CONDITIONS FOR LIQUEFACTION OF GASES

From the Andrew's experiment very useful conditions for the liquefaction of a gas may be stated as :

(i) The temperature must be at or below the critical temperature.

(ii) A suitable pressure must be applied equal to the critical pressure if the temperature is at the critical temperature but becoming less if the temperature is reduced below this value.

van der Waals Equation as an Improvement Over Simple Laws

Van der Waals equation is an improvement over simple laws due to following reasons:

(i) *Van der Waals equation explains the devication from Boyle's law:* Regnault and Amagat studied the behaviour of cases and observed that they do not obey simple gas laws under different conditions of temperature and pressure, more particularly at low temperature and high pressure. They also plotted graphs between the product of pressure and volume ad pressure and found that PV product was not constant as desired by Bole's law. These deviations from ideal behaviour were explained by van der Waals equations.

(ii) *It also explains critical phenomenon.* This equation satisfactorily explains critical phenomenon, observed in the case of gases. The theoretical isotherm (*i.e.*, the curves between V and P at constant T) obtained from the values of V and the experimental values of 'a' and 'b' from van der Waals equation were found to be quite similar. The theoretical isotherms obtained from simple gas laws are quite different from the experimental isotherms.

PRINCIPLE OF CONTINUITY OF STATES

Although, the transition from liquid to vapour and *vice versa* generally takes place suddenly and the two states can be sharply distinguished from each other as can be see from AB and CD parts of 13.1°C isotherm for carbon dioxide yet where occurs a possibility of passing from one state of aggregation to the other without involving sudden change.

Hence for further theoretical considerations it becomes essential to show that the liquid stated does not refer to a sharp and discontinuous transition from gaseous state but rather a continuation of the gaseous phase into the region of very strong intermolecular attractions where the volume becomes very small. The idea of a contiguity from liquid to gaseous state will become evident from the following consideration: The dome-shaped curve depicted b the dashed line connects the end points of the horizontal portions of the isotherms.

As a horizontal portion represents two phases in equilibrium, all points inside the dome correspond to equilibrium between liquid and vapour. For such points it always becomes possible to draw any sharp distinction between liquid and vapour, because there is a definite boundary surface between the two phases. For points lying outside the dome only there is only one phase (liquid or vapour) ad here it is not possible to draw any sharp distinction between liquid and vapour. This fact is termed as the *principle of contiguity of states* and can be demonstrated with reference.

Suppose, that carbon dioxide gas under atmospheric pressure is made to heat a temperature above the critical temperature, as to 48°C and that is then subjected to a pressure greater than critical pressure, as, to 100 atmospheres, the system will change volume as shown by the isotherm 48°C. When the temperature is lowered to 13.1°C while maintaining the pressure at 100 atmospheres, a point on the curve CD will be reached a curve which applies to liquid carbon dioxide. At no point in the process there is any abrupt change of state; the change from gas to liquid will be contiguous. Similarly, it is possible to start with liquid carbon dioxide under the conditions represented, say by the point C.

If we now raise the pressure o the liquid to 10 atmospheres and the temperature to 48°C the condition of the system will be represented by a point on that isotherm; and on reducing the pressure to that of the atmosphere ad then allowing the temperature to fall to 13.1°C, we shall be having carbon dioxide as a gas; and at no time during the whole process any discontinuity of properties will be observed, or an separation into liquid and vapour. The passage from the liquid to the gaseous state would be continuous.

An important conclusion derived from the principle of continuity of states is that one may think of a liquid near the critical point as just a highly compressed gas. If a equation of state can be found which accurately describes the gas close to its critical temperature and pressure that equation should also be able to describe the liquid.

VAN DER WAALS EQUATIONS AND CRITICAL STATE—RELATIONS BETWEEN VAN DER WAALS CONSTANTS AND CRITICAL CONSTANTS

A very good confirmation of the general character of van der Waals equation may be provided b its application to critical phenomenon.

The van der Waals equation may be put as follows:

$$\left(P + \frac{a}{V^2}\right)(V - b) = RT \text{ (putting } = 1) \qquad ...(1)$$

$$PV + \frac{a}{v} - Pb - \frac{ab}{V^2} = RT$$

On multiplying throughout by V^2, we obtain

$$PV^3 + aV - PbV^2 - ab = RTV^2$$

or $$PV^3 + aV - PbV^2 - ab - RTV^2 = 0$$

On arranging in descending powers of V, we get

$$PV^3 - PBV^2 - RTV^2 + aV - ab = 0$$

On dividing b P, we get

$$V^3 - bV^2 - \frac{RTV^2}{P} + \frac{aV}{P} - \frac{ab}{P} = 0$$

$$V^3 - V^2 - \left(b + \frac{RT}{P}\right) + \frac{aV}{P} - \frac{ab}{P} = 0 \qquad ...(2)$$

This is a third degree equation in terms of V (cubic equation in V) and below a certain value of temperature T the equation can be having one of three real roots. With the rise of temperature these roots would approach one another and become identical ultimately. As the temperature gets raised the S-shaped dotted portion of the curve gets reduced to a point. The values for the volume of liquid and the gas approach each other until at this volume of liquid and the gas approach each other until at this point, the three values of V become identical. This point would correspond to the critical temperature T_c and the volume V_c. At the critical point, the three roots of the van der Waals equation would become identical and equal to V_c. But this refers to the value of the three roots of the van der Waals equation at the critical point.

$$(V - V_c)^3 = 0 \qquad ..(3)$$

or

$$V^3 - 3V_cV^2 + 3V_c^2V - V_c^3 = 0 \qquad ...(4)$$

On substituting these values in eq. (2), we get,

$$V^3 - V^2\left(b + \frac{RT_c}{P_c}\right) + \frac{aV}{P_c} - \frac{ab}{P_c} = 0 \qquad ...(5)$$

As both eqs. (4) and (5) are identical then the coefficients of the equal powers of V i the two equations may be equated:

$$3V_c = b + \frac{RT_c}{P_c} \qquad ...(6)$$

$$3V_c^2 = \frac{a}{P_c} \qquad ...(7)$$

$$V_c^3 = \frac{ab}{P_c} \qquad ...(8)$$

On dividing Eq. (8) by Eq. (7), we get,

$$\frac{V_c^3}{3V_c^2} = \frac{ab}{P_c} \times \frac{P_c}{a}$$

or

$$\frac{V_c}{3} = b$$

$$V_c = 3b \qquad ...(9)$$

Substituting the value of V_c from Eq. (9) into Eq. (7), we obtain

$$\frac{a}{P_c} = 3 \times (3b)^2$$

or

$$\frac{a}{P_c} = 27b^2$$

or

$$P_c = \frac{a}{27b^2} \qquad ...(10)$$

Substituting the values of V_c and P_c in equation (6), we get

$$3 \times 3b = b + \frac{RT_c}{a/27b^2}$$

$$9b = b + \frac{RT_c}{a/27b^2}$$

or

$$8b = RT_c \times \frac{27b^2}{a}$$

or

$$T_c = 8b \times \frac{a}{27b^2} \times \frac{1}{R}$$

$$T_c = \frac{8a}{27bR} \qquad ...(11)$$

Derivation of $P_cV_c = \frac{3}{8} RT_c$ from van der Waals Equation and calculation of van der Waals constants 'a' and 'b' in terms of T_c and P_c.

From van der Waals equation, we get

$$V^3 - V^2\left(b + \frac{RT_c}{P_c}\right) + V\frac{a}{P_c} - \frac{ab}{P_c} = 0 \qquad ...(12)$$

and

$$V^3 - 3V_cV^2 + 3V_c^2V - V_c^3 = 0 \qquad ...(13)$$

On equating the coefficients of (12), and (13), we get

$$3V_c = \frac{RT_c}{P_c} + b \qquad ...(14)$$

$$3V_c^2 = \frac{a}{P_c} \qquad ...(15)$$

$$V_c^3 = \frac{ab}{P_c} \qquad ...(16)$$

On dividing Eq. (16) by (15), we get

$$V_c = 3b \quad \text{or} \quad b = \frac{1}{3}V_c \qquad ...(17)$$

On substituting the value of b in (14), we obtain

$$3V_c = \frac{RT_c}{P_c} + \frac{V_c}{3}$$

or

$$3V_c - \frac{V_c}{3} = \frac{RT_c}{P_c}$$

or $$\frac{8}{3}V_c = \frac{RT_c}{P_c}$$

or $$P_cV_c = \frac{3}{8}RT_c \qquad ...(18)$$

On rearranging Eq. (18), we get

$$V_c = \frac{3}{8}\frac{RT_c}{P_c}$$

On squaring, we obtain

$$V_c^2 = \frac{9}{64}\frac{R^2T_c^2}{P_c^2} \qquad ...(19)$$

On substituting the value of $V_c^2 = \frac{a}{3P_c}$ from Eq. (15) into Eq. (19), we get,

$$\frac{a}{3P_c} = \frac{9}{64}\frac{R^2T_c^2}{P_c^2}$$

$$a = \frac{27R^2T_c^2}{64P_c} \qquad ...(20)$$

On substituting the value of $V_c = 3b$ in Eq. (14), we obtain,

$$9b = \frac{RT_c}{P_c} + b$$

or $$8b = \frac{RT_c}{P_c}$$

or $$b = \frac{RT_c}{8P_c} \qquad ...(21)$$

THE PRINCIPLE OF CORRESPONDING STATES

In 1881, van der Waals demonstrated that if the pressure volume and temperature of a gas are expressed in terms of its critical pressure critical volume and critical temperature, then an important generalization called the principle of corresponding states would be obtained.

According to van der Waals equation, we have

$$\left(P + \frac{a}{V^2}\right)(V - b) = RT \qquad ...(1)$$

This equation has constants 'a' and 'b' which cannot be measured directly. Therefore, this equation is not applicable to all gases and in order to make it so, these constants have to be removed.

It has been found experimentally that near the critical point all gases give the same type of isotherms as carbon dioxide, but the actual position of the critical isotherms gets varied with the critical constant of the gas studied. The similarity is great enough to postulate that if for each different gas a suitable scale of pressure, volume and temperature is selected, it will be possible to superimpose all the critical isotherms on one another. If this were done of the critical isotherms it becomes probable that all other isotherms would coincide.

Under such circumstances, there should be one isothermal diagram and one fundamental equation for all gases. Such a scale will be one in which the pressure, volume, and temperature will be expressed as fractions of the critical pressure, volume and temperature.

In order to obtain such an equation, the values of P, V and T must be expressed as corresponding fractions of P_c, V_c and T_c, the critical pressure, volume and temperature,

$$\pi = P/P_c \quad \text{or} \quad P = \pi P_c$$

$$\phi = V/V_c \quad \text{or} \quad V = \phi V_c$$

$$\theta = T/T_c \quad \text{or} \quad T = \theta T_c$$

where π, ϕ and θ are referred to as *reduced pressure, reduced volume and reduced temperature respectively.*

On substituting the values of P, V and T in the van der Waals equation for one mole of an ideal gas, we get

$$\left(\pi P_c + \frac{a}{\phi^2 V_c^2}\right)(\phi V_c - b) = R_q T_c \qquad ...(2)$$

$$P_c = \frac{a}{27b^2}, \quad V_c = 3b$$

$$\text{and} \quad T_c = \frac{8a}{27Rb}$$

On substituting these values in equation (2), we get,

$$\left(\pi \frac{a}{27b^2} + \frac{a}{9\phi^2 b^2}\right)(3\phi\, b - b) = R\theta \frac{8a}{27Rb}$$

or $$\frac{a}{27b^2}\left(\pi + \frac{3}{\phi^2}\right) \times b(3\phi - 1) = \frac{a}{27b}8\theta$$

or $$\frac{a}{27b}\left(\pi + \frac{3}{\phi^2}\right)(3\phi - 1) = \frac{a}{27b}8\theta$$

or $$\left(\pi + \frac{3}{\phi^2}\right)(3\phi - 1) = 8\theta$$

This equation involving neither of the characteristic constants a, b, T_c, P_c and V_c is known as van der Waals reduced equation of state. This equation is applicable to all substances (liquid or gaseous) irrespective of their nature, because it is not involving either of the characteristic constants.

Significance of Law of Corresponding States

It has a great significance in the study of the relationship between physical properties ad chemical constitution of various liquids. As it is known that the pressure has practically no effect on liquids, so while studying the relation between physical properties and chemical constitution the determination of these properties should be performed at the same reduced temperature.

It has been found that the *boiling point of a liquid at atmospheric pressure is almost 2/3rd of its critical temperature* (T_c) *when expressed in absolute scale*. Consequently, boiling point is considered as the *corresponding temperature*. So, various physical properties of liquids such as molecular volumes of various liquids should be compared at their respective boiling points.

In actual practice one cannot use the above reduced equation of state directly. One makes use of the graphs between compressibility factor 'Z' and the reduced pressure at different reduced temperatures. The same graph is applicable to all gases. This can be seen from the following considerations, since $Z = \frac{PV}{RT}$.

Writing this in terms of reduced terms,

We have $$Z = \frac{(\pi P_c)(\phi V_c)}{R(\theta T_c)} = \left(\frac{P_c V_c}{RT_c}\right)\frac{\pi\phi}{\theta} = \frac{3}{8}\frac{\pi\phi}{\theta}$$

Thus, if two gases are having the same reduced temperature and reduced pressure, then their reduced volume will be the same, thus R.H.S.

is independent of the as, thus all gases will have the same value of 'Z'. If isotherms are made to plot in terms of the reduced variables, the same curve should be obtained for all gases.

COMPRESSIBILITY FACTOR

It is known that the equation of state, PV = nRT, derived from the postulates of the kinetic theory, *is valid for an ideal gas only*. Real gases tend to obey this equation only approximately ad that too under conditions of low pressure and high temperature.

The higher the pressure and the lower the temperature, the greater would be the deviations from the ideal behaviour. In general, the most easily liquefiable and highly soluble gases exhibit larger deviations. Hence gases like carbon dioxide, sulphur dioxide and ammonia exhibit much larger deviations than hydrogen, oxygen, nitrogen, etc.

It is best to represent the deviations from ideal behaviour i terms of the compressibility factor, Z, which is defined as

$$Z = \frac{PV}{(PV)_{ideal}} = \frac{PV}{nRT} \qquad ...(1)$$

For an ideal gas, Z = 1 under all conditions of temperature and pressure. The deviation of Z from unity is, therefore a measure of the imperfection of the gas under consideration.

The graphs plotted for the compressibility factors determined for a number of gases over a range of pressures at a constant temperature (0°C). At extremely low pressures, the value of Z for all the gases are close to unity which implies that the gases behave almost ideally. At very high pressures, all the gases have Z more than unity indicating that the gases are less compressible than an ideal gas. This is attributed to the fact that at high pressures, the molecular repulsive forces are dominant.

It moderately low pressures, carbon monoxide, methane and ammonia are more compressible than an ideal gas, *i.e.* $Z < 1$. This is attributed to the fact that at low pressures, the long range attractive forces are dominate and favour compression. The compressibility factor Z goes on decreasing with increase in pressure, passes through a minimum at a certain stage and then starts increasing with increase in pressure. The gases would now becomes less compressible than ideal gas, *i.e.*, $Z > 1$. It can be seen from the figure that although carbon monoxide and methane exhibit marked devications from ideal behaviour only at high pressures, yet ammonia exhibits large deviation even at low pressure.

It can be seen that hydrogen and helium at 0°C are less compressible *at all pressures* than the ideal gas, *i.e.*, $Z > 1$. However, if the temperature is sufficiently low (*e.g.*, below – 165°C for hydrogen and below – 240°C for helium) these gases would also give the same type of Z – P plots as are shown by ammonia, carbon monoxide and methane at 0°C. On the other hand, if the temperature becomes sufficiently high, the Z-P plots of ammonia, carbon monoxide and methane will be almost similar to those of hydrogen and helium at 0°C *i.e.*, the value of Z will increase continuously with increasing pressure.

EXPANSIVITY AND COMPRESSIBILITY

When fluids (gases and liquids) are heated, they expand but the expansion of gases is much more than that of liquids. Similarly, they can be compressed. The variation of volume V with temperature T, keeping pressure P constant is termed as coefficient of thermal expansion or the coefficient of isobaric expansion or simple, expansivity, α of the fluid. Thus.

$$\alpha = \frac{1}{V}\left(\frac{\partial V}{\partial T}\right)_P \quad ...(1)$$

Similarly, the variation of V with P, keeping T constant, is termed *as the coefficient of isothermal compressibility* or, simply, compressibility, β of the fluid, thus,

$$\alpha = -\frac{1}{V}\left(\frac{\partial V}{\partial T}\right)_T \quad ...(2)$$

Generally, as the volume of a gas gets decreased as pressure increases it means that the quantity $(\partial V/\partial P)_T$ is negative. Hence, a minus sign is included in the definition so as to make β positive. It is to be noted that α has the dimensions of T^{-1} and β those of P^{-1}.

Let us derive the value of α and β for an ideal gas.

For n moles of an ideal gas, we have

$$PV = nRT \quad ...(i)$$

or

$$V = nRT/P \quad ...(ii)$$

On differentiating Eq. (ii) with respect to T at constant P, we get,

$$\left(\frac{\partial V}{\partial T}\right)_P = \frac{nR}{P} = \frac{PV}{TP} = \frac{V}{T}$$

$$\therefore \qquad \alpha = \frac{1}{V}\left(\frac{\partial V}{\partial T}\right)_P = \left(\frac{1}{V}\right)\left(\frac{V}{T}\right) = \frac{1}{T}$$

Again, differentiating Eq. (ii) with respect to P at constant T. We get,

$$\left(\frac{\partial V}{\partial P}\right)_T = \frac{nRT}{P^2} = \left(\frac{nRT}{P}\right)\left(-\frac{1}{P}\right) = -\frac{V}{P}$$

$$\therefore \qquad \beta = -\frac{1}{V}\left(\frac{\partial V}{\partial P}\right)_T = \left(-\frac{1}{V}\right)\left(-\frac{V}{P}\right) = \frac{1}{P}.$$

THEOREMS OF PROBABILITY IN STATISTICAL THERMODYNAMIC

The following are the important probability theorems commonly used in statistical thermodynamics:

(1) The number of ways in which N distinguishable particles can he arranged in order is equal to

$$N!$$

(2) The number of different ways in which N particles can be selected from N distinguishable particles irrespective of the order of selection is equal to

$$\frac{N!}{(N-n)!n!}$$

(3) The number of different ways in which n indistinguishable particles can be arranged in g distinguishable states with not more then one particle in each state is equal to

$$\frac{g!}{(g-n)!n!}$$

PROBABILITIES CALCULATION

We can extend the theory of probability in order to include the occurrence of two or more independent events. The combined probability in such cases is given by the product of separate probability of each event. Let us examine some cases:

(i) *Tossing of a coin:* On tossing a single coin, there are two possible aspects that may be presented by a fallen coin, a "head" or a "tail". We may assume here that there is no preference for any particular event, so that after making a large number of throws, the ratio of the number of heads to the number of tails is unity. Thus, for a normal coin the probability of each event is 1/2.

(ii) *Tossing of two unlike coins*: Suppose we toss two coins of different denominations or of different sizes, or of different colours, we have four complexions.

1. $Head_1$ $Head_2$
2. $Head_1$ $Tail_2$
3. $Tail_1$, $Tail_2$
4. $Tail_1$ $Head_2$.

Thus, the probability of any one complexion on a single toss will be 1/4 = 1/2.1/2, Or the composite probability is the product of probability of individual events.

(iii) *Tossing of two similar coins*: If two coins cannot be distinguished from each other, then complexions 2 and 4 of (ii) are indistinguishable whence the probability of one head and one tail is twice than that of two heads or two tails.

(iv) *Tossing of two coins in general*: Let us represent occurrence of heads by A_1, A_2 and the occurrence of tails by B_1, B_2 when two coins are tossed. Then the composite event can be represented by $(A_1 + B_1)(A_2 + B_2) = A_1A_1 + A_1B_2 + A_2B_1 + B_1B_2$.

(v) *Tossing of three unlike coins:* With three unlike coins, we have three distinguishable heads A_1, A_2, A_3 and three distinguishable tails, B_1, B_2, B_3. Here we can have eight equally probable complexions $A_1 A_2 A_3$. $A_1 A_2 B_3$, $A_1 B_2 B_3$, $B_1 A_2 B_3$, $B_1 A_2 B_3$, $B_1 B_2 A_3$ and $B_1 B_2 B_3$ which are really obtained from the product of $(A_1 + B_1)$, $(A_2 + B_2)$, $(A_3 + B_3)$.

(vi) *Tossing of three similar coins:* In this case we have only four probable states with complexions

1. All heads.
2. Two heads and one tail
3. All tails.
4. One head and two tails.

We can tabulate the result as in (iv) for

$(A + B)(A + B)(A + B) = A_3 + 3A^2B + 3AB^2 + B^3$.

(vii) *Tossing of n coins*: Suppose all the coins are different. Then the possible complexions (also known as probability number) will be $(A_1 + B_1)(A_2 + B_2) \ldots\ldots (A_n + B_n)$. And, if the coins are all alike the total number of possible complexions will be

$(A + B)^n = A^n + {}^nC_1A^{n-1}B + {}^nC_2A^{n-2}B^2 + \ldots.{}^nC_rA^{n-r}B^r + \ldots + {}^nC_n$ B^n

$$\text{or} \qquad = \sum_{r=0}^{r=n} C_r A^{n-r}B^r$$

$$\text{or, also } (A + B)^n = \sum_{r=0}^{r=n} C_r A^r B^{n-r}$$

$$\equiv \sum_{r=0}^{r=n} C_r A^{n-r}B^r$$

where ${}^nC_r = \dfrac{n!}{r!(n-r)!}$ ≡ Number of events with r heads/tails.

Now, since A and B indicate one event each, the total number of events are $(1 + 1)^n = 2^n = {}^nC_1 + {}^nC_2 + \ldots + {}^nC_n$

∴ The probability of r heads and (n – r) tails to appear is

$$P(r, n-r) = \frac{{}^nC_r}{2^n} = \frac{n!}{r!(n-r)!}.\frac{1}{2^n}$$

Maximum and minimum probability: To find out which of the combination of heads and tails is most likely in a large number of throws of n coins, we have to find the maximum value of ${}^nC_r$. It is known from algebra that the value of ${}^nC_r$ is greatest for r = n/2 when n is even and r = n+1/2 when n is odd.

$$P_{max} = \frac{{}^nC_{n/2}}{2^n} = \frac{n!}{(n/2)!(n/2)}.\frac{1}{2^n}$$

Similarly, the minimum probability, P_{min} obtained for r = 0 or r = n as ${}^nC_n = 1$.

whence $P_{min} = 1/2^n$

(viii) *Tossing of a single coin n times*: When we toss the same coin second, third,... nth time, its behaviour is again unpredictable. It may show a head or a tail each time with 1/2 probability. Hence a single coin tossed n times (and of course, making large number of such attempts, if we want to have satisfactory results) is just equivalent to the case of tossing of n coins as discussed in (vii).

Another equivalent illustration can be: Take 2 boxes, identical in all respects, and throw n balls from a great distance. Assuming that each ball will fall into one or the other box, the distribution of balls in the boxes will be governed by the relation (vii),

(ix) *Tossing of a dice*: If we toss of dice, having six faces, the probability for each face showing up will be 1/6. Since now the total number of aspects is 6 and only one can show up at a time with equal probability. Again, we may take 2, 3,.... n dices and throw them and calculate the composite probability for a certain complexion.

(x) *Tossing of n dices each having m faces*: Here let us assume that dice 1 has faces a_1, a_2, a_3... a_m; dice 2 has faces b_1 b_2 b_3 ... b_m; dice 3 has faces c_1, c_2, c_3 ; etc., then the possible complexions with n dices will be the sum of each aspect

a_1, a_2, a_3... a_m, v_1, b_2, b_3 ... b_m , c_1, c_2, c_3 ... n sets and, if the dice are all alike, we have

$m_{.a}$, $m_{.b}$, $m_{.c}$... n sets.

If the probability for occurring each of the face is same, *i.e.*, a = b = c... = one aspect, we get total number of possible aspects

$= m\ m\ m \ldots n \text{ times} = m^n$

One can show that the most probable distribution corresponds to equitable distribution of faces *i.e.*, face 1, face 2, face 3, etc., will appear in equal number. Then all the dices showing face 1 have the probability of only $1/m^n$.

A similar illustration to this is : Take *m identical boxes and throw n* balls onto them from a great distance. We will take up this problem later in more detail. As in actual analysis of physical problems we have a large number of particles which are distributed into a large number of (hypothetical) boxes.

(xi) *Probability with weightage*: So far we have considered the case in which the probability of occurring an event and its failure is the same. But there can be cases where the probability of occurring a particular event is greater than that for any other event.

Let a be the probability of occurring an event and b be the probability of its failure, per trial. Then, if the trial is repeated n times, the probability of r favourable events is

$a^r.b^{n-r} = a^r (1 - a)^{n-r}$

But there are ${}^nC_r$ ways for r favourable events, therefore, the probability of occurring r events is

$$P(r, n - r) = {}^nC_r\, a^r (1 - a)^{n-r}$$

Thus, for example, in a dice a = 1/6

$$\therefore \quad P(r, n - r) = {}^nC_r \left(\frac{1}{6}\right)^r \left(\frac{5}{6}\right)^{n-r}$$

But, if certain number, say 3, has a = 2/7 then in n tosses the number 3 will occur r times with a probability.

(xii) *Postulate of equal a priori probability:* It states that all accessible microstates corresponding to a macrostate are equally probable. Thus, the postulate claims that the probability of finding a molecule in one region of phase space is identical with that for any other region of equal volume provided the regions are obeying other conditions of equality (*i.e.*, the regions have extensions of same magnitude and correspond to the same energy). Thus, it follows that the probability of occurrence of a macrostate is proportional to the number of microstates associated with it

i.e., $P_i \propto \Omega_i$

where P_i is the probability and Ω_i is the total number of accessible microstates, in the ith set.

ACCESSIBLE MICROSTATES

Let us now count the number of energy states available in a macrosystem. Let E be the total energy of this system which has been divided into small energy intervals δ E so that a sufficient number of quantum states are available in each interval of δ E. Now, if Ω (E) is the number density of such states in the energy interval E and E + δE then

$$\Omega(E) = \rho(E)\, \delta E \qquad \text{...(i)}$$

where ρ (E) is density of states (the number of states in unit energy interval). Thus

$$\Omega(E) \propto \delta E \qquad \text{...(ii)}$$

Let the value of Ω(E) changes with the magnitude of energy interval. Hence we may consider Ω(E) as a smoothly varying function of E.

Again, if $\phi(E)$ denotes the number of states with energy up to E and $\phi(E + \delta E)$ denotes the number of states with energy up to $E + \delta E$ then

$$\Omega(E) = \phi(E + \delta E) - \phi(E) \equiv \frac{\delta\phi}{\delta E}\delta E \qquad ...(iii)$$

Comment on ρ (E). Relations (i) and (ii) (possibly) indicate that $\rho(E)$ is a constant. But it is not so. In fact ρ (E) is an increasing function of E. Let us see how! We know that the total number of microstates is equal to the product of thermodynamic probabilities of different macrostates (composite probability is obtained, always by multiplying probabilities for different complexions), *i.e.*,

$$\Omega = \Omega_1 \, \Omega_2 \, \Omega_3 \ldots \Omega_i \ldots \Omega_f$$

or $$\Omega = \prod_1 \Omega_i \; (i = 1 \text{ to } f)$$

where f is the degree of freedom.

But Ω_i should be somehow related to E, *i.e.*,

$$\Omega \propto E^a \quad \therefore \Omega \propto E^{fa} \qquad ...(iv)$$

usually $a \cong 1/2$ and/ranges between 10^{22} and 10^{28}. This suggests that for a certain energy interval δE, $\rho(E)$ increases with the value of E.

STATISTICAL EQUILIBRIUM

Consider that a thermodynamical isolated system consists of N particles. The energy states available to the particles are E_1, E_2, E_3 etc. These energy states may be quantized or may be continuous and are due to vibrational and rotational energy of the particles.

Suppose, that at any given instant of time, n_1 particles are in state of energy E_1, n_2, particles with energy state E_2 and so on.

The total number of particles in the system is,

$$N = n_1 + n_2 + n_3 + \ldots\ldots = \sum_i n_i \qquad ...(i)$$

Here $i = 1, 2, 3$ etc.

The total energy of the system,

$$U = n_1E_1 + n_2E_2 + n_3E_3 + \ldots$$

$$U = \sum_i n_i E_i \qquad ...(ii)$$

Equation (ii) refers to the total energy of a system in which the particles are non-interacting. Here, the energy of each particle depends only on the coordinates of the particle in the system. For an isolated system, the total energy U is constant.

For isolated system.

$$U = \sum_i n_i E_i = \text{constant} \qquad \text{...(iii)}$$

Consider a gas having N molecules at a certain temperature and pressure. Its volume, temperature and pressure are kept constant *i.e.*, the system is isolated. The total energy of this system remains constant. But, the molecules of the gas collide with each other and also with the walls of the container. Consequently, the number of molecules change from one energy state to the other energy state. It means that the value of n_1, n_2, n_3 etc., continuously change.

It can be reasonably assumed that for each microscopic State of a system of particles, there is a particular most favoured distribution. When this distribution or partition is reached, the system attains statistical equilibrium. For an isolated system, the values of n_1, n_2, n_3 etc., vary only near the values corresponding to the most probable distribution. Hence, the basic problem in statistical thermodynamics is to obtain the most probable distribution law for a given composition of the system.

In practice, three most probable distribution laws are used. They are:

(i) Maxwell-Boltzmann Distribution Law.

(ii) Fermi-Dirac Distribution Law (Statistics).

(iii) Bose-Einstein Distribution Law (Statistics).

MAXWELL-BOLTZMANN DISTRIBUTION LAW

Consider a system that contains a large number of particles that are identical and distinguishable. The identical particles refer to the particles having the same structure. The distinction between the particles is due to their energy states at a given instant. Four particles are in energy state E_1, 2 in state E_2, zero in state E_3 and so on. It is assumed that all the energy states are accessible to each particle. Consequently, it can be assumed that the probability of any particular partition is proportional to the number of different ways in which the particles can be distributed in the existing available energy states so as to produce the desired partition.

To obtain the required distribution as the first particle a in state E_1 can be selected in N ways. The second particle b in state E_1 can be selected in (N – 1) ways and so on. Therefore, the total number of ways in which the first four particles in state E_1 can be selected is given by,

$$N(N-1)(N-2)(N-3) = \frac{N!}{(N-4)!}$$

Moreover, the four particles in state E_1 can be arranged in 4! different orders. For example, *abcd, bcda, cdab* and so on. There are 24 ways. But, it is immaterial for these particles to be arranged in any particular order in state E_1, because they are identical. Thus, the total number of *distinguishable* different ways are,

$$\frac{N!}{4!(N-4)!}$$

In general, if the first state consists of n_1 particles, the distinguishable different ways, for arranging n_1 particles in state E_1, are,

$$p_1 = \frac{N!}{n_1!(N-n_1)!} \qquad \text{...(i)}$$

For the second state E_2, only $(N - n_1)$ particles are available and n_2 particles are in state E_2. The number of distinguishable different ways are,

$$p_2 = \frac{(N-n_1)!}{n_2!(N-n_1-n_2)!} \qquad \text{...(ii)}$$

If this process is continued for all the available states, the total number of distinguishable ways are obtained by multiplying P_1,P_2, P_3 etc.

$$P = p_1 \times p_2 \times p_3 \times \text{......}$$

$$P = \left[\frac{N!}{n_1!(N-n_1)!}\right]\left[\frac{(N-n_1)!}{n_2!(N-n_1-n_2)}\right] \times \text{....}$$

$$P = \frac{N!}{n_1!n_2!n_3!...} \qquad \text{...(iii)}$$

The distinguishable ways are,

$$P = \frac{N!}{4!2!0!3!1!} \qquad \text{...(iv)}$$

Here 0! is equal to one.

It has been assumed so far that all the available states have the same probability of occupation by the particles. However, it may happen that the states have different intrinsic probabilities say g_i.

For example, a particular energy state may be favourable with more different angular momentum states than the rest and hence it is more likely to be occupied. Taking into account this intrinsic probability factor, the value of P will be different.

If g_i is the probability of locating a particle in a certain energy state E_i, then the probability of locating 2 particles in the same state is $g_i \times g_i = g^2_i$. For n_i particles, the probability is $g_i n_i$. Hence the total probability for a given distribution is given by

$$P = \frac{N! g_1 n_1 g_2 n_2 g_3 n_3 \ldots}{n_1! n_2! n_3! n_4! \ldots} \quad \ldots(v)$$

Here n_1, n_2, n_3 etc., are the number of particles in states E_1, E_2, E_3, etc. and g_1, g_2, g_3 etc., are the intrinsic probabilities for states E_1, E_2, E_3 etc.

If all the particles are further assumed to be indistinguishable *i.e.*, particles in state E, and particles in state E_4 cannot be distinguished, then N! permutations among the particles themselves and occupying the different states result in the same distribution. The probability in this case is given by

$$P = \frac{1}{N!}\left[\frac{N! g_1 n_1 g_2 n_2 g_3 n_3 \ldots}{n_1! n_2! n_3! \ldots}\right]$$

$$\therefore \quad P = \frac{g_1 n_1 g_2 n_2 g_3 n_3 \ldots}{n_1! n_2! n_3! \ldots}$$

$$\therefore \quad P = \prod_i^N \frac{g_i n_i}{n_1!} \quad \ldots(vi)$$

Here Π is the product sign (since probability is the product of such distributions.

The most probable distribution can be obtained by evaluating the maximum value of $\log_{e}$, P in equation (vi). This should also satisfy the two conditions that

$$\sum_i n_i = N \quad \ldots(vii)$$

and

$$\sum_i n_i \; E_i = U \quad \ldots(vii)$$

According to Stirling's approximation

$$\log_e x! = x \log_e x - x$$

From equation (vi), applying Stirling's approximation, we get

$$\log_e P = \sum_i (n_i \log_e g_i - \log_e n_i!)$$

$$= \sum_i [n_i \log_e g_i - (n_i \log_e n_i - n_i)]$$

$$= \sum_i (n_i \log_e g_i - n_i \log_e n_i + n_i)$$

$$= \sum_i n_i - \sum_i (n_i \log_e n_i - n_i \log_e g_i)$$

$$\log_e P = N \sum_i n_i \log_e \frac{n_i}{g_i} \qquad ...(ix)$$

Differentiating equation (ix)

$$d(\log_e P) = \sum_i (dn_i) \log_e \frac{n_i}{g_i} - \sum_i n_i d\left(\log_e \frac{n_i}{g_i}\right)$$

$$= \sum_i dn_i \log_e \frac{n_i}{g_i} - \sum_i n_i \frac{dn_i}{n_i}$$

$$= \sum_i dn_i \log_e \frac{n_i}{g_i} - \sum_i dn_i$$

But $\sum_i dn_i = 0.$

$$\therefore \quad -d(\log_e P) = \sum_i \left[\log_e\left(\frac{n_i}{g_i}\right)\right] dn_i \qquad ...(x)$$

To obtain the maximum value of P,

$$d(\log_e P) = 0 \qquad ...(xi)$$

But $\Sigma dn_i = 0$ ...(xii)

and $\Sigma E_i dn_i = 0$ ...(xiii)

Multiplying (xii) by α and (xiii) by β and adding to equation (xi), we get

$$\sum_i \left[\log_e\left(\frac{n_i}{g_i}\right) + \alpha + \beta E_i\right] dn_i = 0$$

The equilibrium distribution is possible if

$$\log_e \frac{n_i}{g_i} + \alpha + \beta\, Ei = 0 \qquad ...(xiv)$$

or $$\frac{n_i}{g_i} = e^{-\alpha-\beta Ei}$$

or $$n_i = g_i\ e^{-\alpha-\beta Ei} \qquad ...(xv)$$

This gives the maximum probability distribution and a and p are two parameters that depend upon the physical property of the system.

$$N = n_1 + n_2 + n_3 + ...$$

$$N = g_1 e^{-\alpha-\beta E1} + g_1 e^{-\alpha-\beta E2} + ...$$

$$N = e^{-\alpha}[g_1 e^{-\beta E1} + g_2 e^{-\beta E2} + ...]$$

$$N = e^{-\alpha}[\sum_i g_i\ e^{-\beta Ei}]$$

Take $[\sum_i g_i\ e^{-\beta Ei} = Z$ ...(xvi)

Here, Z is called the partition function.

$$N = e^{-\alpha}(Z) \qquad ....(xvii)$$

$$e^{-\alpha} = \frac{N}{Z}$$

Substituting this value in equation (vii)

$$n_i = \left(\frac{N}{Z}\right)[g_i e^{-\beta Ei}] \qquad ....(xviii)$$

This equation refers to Maxwell-Boltzmann Distribution Law.

MAXWELL-BOLTZMANN DISTRIBUTION IN TERMS OF TEMPERATURE

The total energy of an isolated system is given by

$$U = \sum_i n_i\ E_i$$

$$U = n_1 E_1 + n_2 E_2 + n_3 E_3 + ...$$

$$U = (g_1 e^{-\alpha-\beta E1})\ E_1 + (g_2 e^{-\alpha-\beta E2})\ E_2 + ...$$

$$U = e^{-\alpha}[g_1 E_1 e^{-\beta E1} + g_2 E_2 e^{-\beta E2} + ...]$$

But $$e^{-\alpha} = \frac{N}{Z}$$

$\therefore$ $$U = \frac{N}{Z}[g_1 E_1 e^{-\beta E1} + g_2 E_2 e^{-\beta E2} + ...]$$

$$U = \frac{N}{Z}\sum_i g_i\ E_i e^{-\beta Ei}$$

Here $$Z = \sum_i g_i e^{-\beta E_i} \quad ...(i)$$

$$\frac{dZ}{d\beta} = \frac{d}{d\beta}\sum_i g_i e^{-\beta E_i} = -\sum_i g_i E_i e^{-\beta E_i}$$

$$\therefore \quad \sum_i g_i E_i e^{-\beta E_i} = \frac{d}{d\beta}\sum_i g_i E_i e^{-\beta E_i}$$

Substituting this value in equation (i)

$$U = -\frac{N}{Z}\frac{d}{d\beta}\sum_i g_i e^{-\beta E_i}$$

$$U = -\frac{N}{Z}\frac{d}{d\beta}$$

$$U = -N\frac{d}{d\beta}[\log_e Z] \quad ...(ii)$$

The average energy of a particle is

$$E_{av} = -\frac{U}{N} = -\frac{d}{d\beta}(\log_e Z) \quad ...(iii)$$

This shows that for a given system, the total energy U, the partition function Z, and the average energy of the particle E_{av}, depend on the parameter β.

Therefore, β may be taken to characterise the internal energy of the system and has the units per joule. If T is the temperature in degrees kelvin, then it is more convenient to represent parameter β as

$$\beta = \frac{1}{kT} \text{ or } KT = \frac{1}{\beta} \quad ...(iv)$$

Here kT has the unit of energy *i.e.*, joule and k is the Boltzmann's constant. Its units are J/K. The value of k is given by

$$k = 1.3805 \times 10^{-23} \text{ J/K}$$

Substituting the value of

$$\beta = \frac{1}{kT}$$

in all the equations, we get

$$Z = \sum_i g_i e^{-(Ei/kT)} \quad ...(v)$$

$$n_i = \frac{N}{Z}\sum_i g_i e^{-(Ei/kT)} \quad ...(vi)$$

This equation represents the Maxwell-Boltzmann distribution law in terms of the temperature of the system.

As $\beta = \frac{1}{kT}$

$$d\beta = -\frac{dT}{kT^2}$$

Substituting this value in equation (ii)

$$\therefore \quad U = (kNT^2)\frac{d}{dT}(\log_e Z) \qquad ...(vii)$$

Also, from equation (iii)

$$E_{av} = (kT^2)\frac{d}{dT}(\log_e Z) \qquad ...(viii)$$

Equation (viii) gives the relation between the average energy of the particle and its temperature under equilibrium position. Hence, the temperature of a system in statistical equilibrium is the physical quantity related to the average energy of the particle of the system.

QUANTUM STATISTICS

Maxwell-Boltzmann distribution, it has been assumed that all the energy levels are accessible to all the particles of the system. However, there may be certain levels prohibited to a certain group of particles. Each energy state is associated with a certain available wave function. The probability of a particular distribution is restricted by the available wave functions of any state. These restrictions are taken into account in quantum statistics. There are two types of quantum statistics

(i) Fermi-Dirac statistics

(ii) Bose-Einstein statistics.

In Fermi-Dirac statistics the particles are assumed to obey Pauli's exclusion principle. They are characterised by antisymmetric wave functions. The particles under, this category are called *fermions*. Protons neutrons and electrons are fermions.

In the case of Bose-Einstein statistics, it is assumed that the particles are not restricted by Pauli's exclusion principle and are characterised by symmetric wave functions. These particles are called *bosons*. It has been experimentally found that all particles having spin zero or integral multiple of 1 are bosons. Helium nuclei and mesons are bosons.

In both the kinds of quantum statistics, the particles are identical and indistinguishable. At high temperature and low pressure, all the three statistics give practically the same result.

FERMI-DIRAC DISTRIBUTION LAW

Consider a system consisting of a large number of particles. It is assumed:

(i) The particles are *identical* and *indistinguishable*, and

(ii) The particles obey exclusion principle. It means that no two particles can have the same dynamical state and the wave function of the whole system must be antisymmetric. The particle satisfying these conditions are called *fermions*. In general, all fundamental particles with spin are fermions.

In quantum statistics the intrinsic probability g_i is governed by the different quantum states relating to a given energy *i.e.*, the degeneracy of the energy state. Each quantum state corresponds to a particle wave function. The Wave functions are determined by each of the possible arrangements of quantum number corresponding to a given energy level. For particles with spin 1/2, in the absence of magnetic forces, each of the particle may be in the energy states with spin +1/2 or –1/2. Hence the intrinsic probability g_i in this case is 2.

For motion in a central field, the energy of the particle is independent of the orientation of the orbital angular momentum Due to this, a degeneracy of $(2\,l + 1)$ is introduced and it is the value of g for that particular energy state. If the particles possess spin, the total degeneracy = $2\,(2l + 1)$. As no two particles can be in the same energy state having the same quantum number, the instrinsic probabilities g_i's give the maximum number of particles that can be accommodated in a particular energy level without violating the exclusion principle *i.e.*,

$$n_i \leq g_i$$

It means that the n_i value for a given distribution does not exceed the corresponding value of g_i.

Let n_i, be the number of particles for energy level E_i The first particle can be placed in any one of the available g_i states *i.e.*, this particle can be assigned to any of the gi sets of quantum numbers. Thus, the first particle can be distributed in g_i different ways. Similarly, the second particle can be arranged in $(g_i - 1)$ different ways and the process continues.

Thus, the total number of different ways of arranging n_i particles among the available g_i states with energy level E_i is

$$= g_i(g_i - 1)(g_i - 2) \ldots\ldots[g_i - (n_i - 1)]$$

$$= \frac{g_i!}{(g_i - n_i)!} \qquad \ldots(i)$$

Further, if the particles are taken to be indistinguishable, it will not be possible to detect any difference when n_i particles are reshuffled into different states occupied by them in the energy level E_i. Therefore, the total number of different and distinguishable ways is,

$$= \frac{g_i!}{n_i!(g_i - n_i)!} \qquad \ldots(ii)$$

Therefore, the total number of different and distinguishable ways of getting the distribution n_1, n_2, n_3, etc., among the various energy levels, E_1, E_2, E_3...,etc.. can be obtained by multiplying the various factors.

$$\therefore \qquad P = \frac{g_1!}{n_1!(g_1 - n_1)!} \cdot \frac{g_{2i}!}{n_2!(g_2 - n_2)!} \ldots$$

$$P = \prod_i \frac{g_i!}{n_i!(g_i - n_i)!} \qquad \ldots(iii)$$

The most probable distribution can by obtained by evaluating the maximum value of $\log_e$, P in equation (iii).

This should also satisfy the condition that

$$\sum_i n_i = N$$

and $\sum_i n_i E_i = U$.

According to Stirling's approximation

$$\log_e x! = x \log_e x - x$$

From equation (iii), applying Stirling's approximation,

$$\log_e P = \sum_i [(g_i \log_e g_i - g_1) - (n_i \log_e n_i - n_i)$$
$$- [(g_i - n_1) \log_e (g_i - n_1) - (g_i - n_1)]$$

$$\log_e P = \sum_i [g_i \log_e g_i - n_1 - \log_e n_i - (g_i - n_i)$$

$$\log_e (g_i - n_1)] \quad ....(iv)$$

Differentiating equation (iv)

$$-d(\log_e P) = \sum_i [\log_e n_i - \log_e (g_i - n_i)]dn_i$$

To obtain the maximum value of P,

$$d(\log_e P) = 0$$

$$\therefore \quad \sum_i [\log_e n_i - \log_e (g_i - n_i)]dn_i = 0 \quad ...(v)$$

But $$\sum_i dn_i = 0 \quad ...(vi)$$

and $$\sum_i E_i dn_i = 0 \quad ...(vii)$$

Multiplying (vi) by α and (vii) by β and adding to equation (v), we get

$$\sum_i [\log_e n_i - \log_e (g_i - n_1) + \alpha + \beta E_i]dn_i = 0$$

The equilibrium distribution is possible if

$$\therefore \quad \log_e n_i - \log_e (g_i - n_1) + \alpha + \beta E_i = 0 \quad ...(viii)$$

$$\log_e\left(\frac{n_i}{g_i - n_i}\right) = -\alpha + \beta E_i$$

$$\frac{n_i}{g_i - n_i} = e^{-\alpha - \beta Ei}$$

$$\frac{g_i - n_i}{n_i} = e^{\alpha + \beta Ei}$$

$$\frac{g_i}{n_i} - 1 = e^{\alpha + \beta Ei}$$

$$\frac{g_i}{n_i} = (e^{\alpha + \beta Ei}) + 1$$

$$n_i = \left(\frac{g_i}{e^{\alpha + \beta E_i}}\right) + 1 \quad ...(ix)$$

Equation (ix) represents the Fermi-Dirac distribution law. The parameter has the same role as in the case of Maxwell-Boltzmann distribution law *i.e.*, for the system consisting of fermions in statistical equilibrium.

$$\beta = \frac{1}{kT}$$

$$\therefore \qquad n_i = \frac{g_i}{[e^{(\alpha + E_i/kT)}] + 1}$$

In most cases the value of a is negative and is taken to be equal to

$$\frac{-E_F}{kT}$$

$$n_i = \frac{g_i}{\left[e^{(E_i - E_F)/kT)}\right] + 1}$$

The value of E_F is positive and is independent of temperature.

For T = 0, all the energy states are fully occupied and $n_i = g_i$ All the states with $E > E_F$, are empty *i.e.*, $n_i = 0$.

$$\left[e^{(E_i - E_F)/kT)}\right] = \begin{cases} 0 \text{ for } E_i - E_F < 0 \\ \infty \text{ for } E_i - E_F > 0 \end{cases}$$

In the case of Fermi-Dirac statistics, the accumulation of particles at the ground level is not allowed and at temperature T = 0, the particles occupy the lowest energy levels upto E_F. Here, the energy E_F gives the indication of the maximum energy of the fermions in the system. By is also called Fermi energy. For higher temperatures, the particles, occupy higher energy states greater than E_F. The curves indicate that only those fermions with energies close to E_F can move into unoccupied higher energy states. If $k\theta_F = E_F$, the temperature θ_F is called the Fermi temperature.

MAXWELL-BOLTZMANN DISTRIBUTION AND IDEAL GAS

Most of the gases obey Maxwell-Boltzmann distribution over a wide range of temperature. Consider a gas having mono-atomic molecules. It is assumed that the gas is ideal and has only kinetic energy of translation. Intermolecular attraction is assumed to be absent. If the kinetic energy is not quantized but considered to possess continuous values of energy, then the partition function Z can be written as

$$Z = \int_0^{\infty} e^{-(E/kT)}\, g(E)dE$$

Here g_i is replaced by g (E) dE

But $$g(E)dE = \left(\frac{4\pi V(2m^3)^{1/2}}{h^3}\right)E^{1/2}dE$$

Here V is the volume occupied by the gas and h is Planck's constant.

$$Z = \frac{4pV(2m^3)^{1/2}}{h^3}\int_0^\infty E^{1/2}e^{-(E/kT)}dE$$

$$Z = \frac{4\pi V(2m^3)^{1/2}}{h^3}\left[\frac{\sqrt{\pi(kT)^3}}{2}\right]$$

$$Z = \frac{V(2\pi mkT)^{3/2}}{h^3} \qquad ...(i)$$

Equation (i) represents the partition function for an ideal mono-atomic gas in terms of the volume and temperature of the gas.

Taking logarithms on both sides

$$\log_e Z = \log_e\left[\frac{(2\pi m)^{3/2}V}{h^3}\right] + \log_e (kT)^{3/2}$$

$$\log_e Z = C + \frac{3}{2}\log_e kT \qquad ...(ii)$$

Also $$E_{av} = (kT^2)\frac{d}{dT}(\log_e Z)$$

$$E_{av} = (kT^2)\frac{d}{dT}\left[C + \frac{3}{2}\log_e kT\right]$$

$$= (kT^2)\frac{d}{dT}\left[C + \frac{3}{2}\log_e k + \frac{3}{2}\log_e T\right]$$

$$E_{av} = \frac{3}{2}kT \qquad ...(iii)$$

The total energy,

$$U = NE_{av}$$

$$U = N\left(\frac{3}{2}kT\right)$$

$$U = \frac{3}{2}NkT. \qquad ...(iv)$$

Equation (iv) shows that the internal energy of an ideal monoatomic gas depends only on its temperature. The same relation does not however, hold good for real gases. In the case of real gases the internal energy is partly potential and partly kinetic. The total energy depends on the volume of the gas.

ELECTRON GAS

Electrons in a metal belong to a most characteristic system of fermions because electrons obey the exclusion principle. For electrons in a metal, the energy levels are grouped in bands. Practically at all temperatures, the lower level energy bands are filled with electrons. The upper level energy bands are only partially filled with electrons. The distribution of electrons is to be considered only in the upper bands called the conduction band.

The zero energy level is taken at the lowest level of the conduction band. It is also assumed that the electrons have free movement within the conductor, provided the energy associated with the electrons is of the order of upper level energy bands.

As the energy of the electron in the conduction band is continuous, the term g_i is replaced by g(E) dE. Here dn electrons have energy in the range E and E + dE.. According to Fermi-Dirac distribution law

$$n_i = \frac{g_i}{\left[e^{(E_i - E_F)/kT)}\right] + 1} \qquad \text{...(i)}$$

Substituting the value of g_i = g (E) dB in equation (i) and replacing n_i by dn

$$dn = \frac{g(E)dE}{\left[e^{(E_i - E_F)/kT)}\right] + 1}$$

As an electron has the spin ±1/2, the total number of states in the sphere is twice. $V/(2\pi)^3$ refers to the translational states per unit volume in the Fermi space Fermi sphere of radius kF has the total number of particles accommodated,

$$n = \frac{2\left[\frac{V}{(2\pi)^3}\right]\left(\frac{4}{3}\pi(kF)^3\right)}{\left[e^{(E_i - E_F)/kT)}\right] + 1}$$

Here $$E = \frac{p^2}{2m} \frac{\left(\frac{h}{2\pi}\right)^2 (kF)^2}{2m}$$

$$\therefore \qquad kF = \frac{p}{h/2\pi} = \frac{2\pi p}{h}$$

$$\therefore \qquad n = \frac{2\left[\frac{V}{(2\pi)^3}\right]\left(\frac{4}{3}\pi\left(\frac{2\pi p}{h}\right)^3\right)}{\left[e^{(E_i - E_F)/kT)}\right] + 1}$$

But $p = (2mE)^{1/2}$

$$\therefore \qquad n = \frac{2\left[\frac{V}{(2\pi)^3}\right]\left[\left(\frac{4}{3}\pi\left(\frac{2\pi p}{h}\right)^3 (2mE)^{3/2}\right)\right]}{\left[e^{(E_i - E_F)/kT)}\right] + 1}$$

$$n = \frac{\left(\frac{8\pi V}{(3h^3}\right)(2m)((2m)^{1/2}E^{3/2}}{\left[e^{(E_i - E_F)/kT)}\right] + 1}$$

Differentiating

$$dn = \frac{\left(\frac{8\pi V}{(3h^3}\right)(2m)((2m)^{1/2} \times \frac{3}{2}E^{1/2}dE}{\left[e^{(E_i - E_F)/kT)}\right] + 1}$$

$$\frac{dn}{dE} = \frac{\left[\frac{8\pi V(2m^3)^{1/2}}{(3h^3}\right]E^{1/2}}{\left[e^{(E_i - E_F)/kT)}\right] + 1} \qquad \text{...(ii)}$$

Equation (ii) represents the energy distribution for free electrons. This is also called Fermi-Dirac formula of free fermions.

At $\qquad T = 0K,$

$$p = (2m\ E_F)^{1/2} \qquad \text{...(iii)}$$

and $N = \left(\frac{2V}{h^3}\right)\left(\frac{4}{3}\pi p^3\right)$

or $$P = \left(\frac{3Nh^3}{8\pi V}\right)^{1/3} \qquad \text{...(iv)}$$

Equating (iii) and (iv)

$$(2mE_F)^{1/2} = \left(\frac{3Nh^3}{8\pi V}\right)^{1/3}$$

Squaring $2mE_F)^{1/2} = \left(\frac{3Nh^3}{8\pi V}\right)^{2/3}$

or $$E_F = \frac{h^2}{8m}\left(\frac{3N}{\pi V}\right)^{2/3} \quad ...(v)$$

Knowing the value of N/V *i.e.*, the number of free electrons per unit volume, the Fermi energy for electrons in a metal can be obtained.

For silver, the value $\frac{N}{V}$

$$= 5.86 \times 10^{28} \text{ electrons/m}^3$$

$$\therefore \quad E_F = \left(\frac{(6.624\times10^{-34})^2}{8\times9\times10^{-31}}\right)\left(\frac{3\times5.86\times10^{28}}{3.14}\right)^{2/3}$$

$$E_F = 9 \times 10^{-19}\text{J}.$$

But $1\text{eV} = 1.6 \times 10^{-19}\text{J}$

$$\therefore \quad E_F = \frac{9\times10^{-19}}{1.6\times10^{-19}}$$

$$E_F = 5.625 \text{ eV}.$$

It means that the maximum kinetic energy of free electrons in silver at absolute zero temperature is 5.6 eV.

The Fermi temperature θ_F is given by

$$\theta_F = \frac{E_F}{k}$$

As E_F practically independent of temperature, the value θ_F is fixed for a given metal. The values of θ_F and E_F are given in the following table.

Metals	***E_F (electron Vol.ts)***	***θ_F(K)***
Potassium	2.14	2.4×10^4
Sodium	312	3.7×10^4
Lithium	4.72	6.5×10^4
Silver	5.51	6.4×10^4
Gold	5.64	6.4×10^4
Copper	7.04	8.2×10^4

DISTRIBUTION LAW OF BOSE-EINSTEIN

Bose-Einstein distribution is applied to systems composed of identical and indistinguishable particles that are not restricted by the exclusion principle. In such systems, there is no limit to the number of particles occupying a particular quantum state. These particles are called *bosons*. Their spin is zero or 1. Mesons and helium nuclei are examples of bosons.

The Bose-Einstein statistics values of g_i refer to degeneracy of each energy level. Suppose, that n_i particles are arranged in a row and distributed among g_i quantum states with $(g_i - 1)$ partitions in between.

The total number of possible arrangements of particles and partitions is equal to the total number of permutations of $(n_i + g_i - 1)$ objects in a row. Therefore the total possible ways of arranging n_i particles with $g_i - 1$ partitions

$$= (n_i + g_i - 1)!$$

As the particles are identical and indistinguishable the possible number of distinct arrangements

$$= \frac{(n_i + g_i - 1)!}{n_i!(g_i - 1)!}$$

The total number of distinguishable and distinct ways of arranging N particles in all the available energy states is given by

$$P = \frac{(n_1 + g_1 - 1)!}{n_1!(g_1 - 1)!} \times \frac{(n_2 + g_2 - 1)!}{n_2!(g_2 - 1)!} \times \ldots$$

$$P = \prod_i \frac{(n_i + g_i - 1)!}{n_i!(g_i - 1)!} \quad \text{...(i)}$$

The most probable distribution can be obtained by finding the maximum value of $\log_e$, P.

According to Stirling's approximation

$$\log_e x! = x \log_e x - x$$

Also $$\sum_i n_i = N \quad \text{...(ii)}$$

and $$\sum_i n_i E_i = U \quad \text{...(iii)}$$

From equation (i), applying Stirling's approximation

$$\log_e P = \sum_i [\log_e (n_i + g_i - 1) - \log_e n_i! \log_e (g_i - 1)!]$$

$$\log_e P = \sum_i [(n_i + g_i - 1)\log_e(n_i + g_i - 1) - (n_i + g_i - 1)$$
$$- (n_i \log_e n_i - n_i) - (g_i - 1)\log_e(g_i - 1) - (g_i - 1)]$$

$$\log_e P = \sum_i [(n_i + g_i - 1)\log_e(n_i + g_i - 1)$$
$$- n_i \log_e - n_i \;\; -(g_i - 1)\log_e(g_i - 1)] \quad ...(iv)$$

The maximum value of P is obtained by taking

$$d(\log_e P) = 0.$$

Differentiating equation P.

$$d(\log_e P) = \sum_i [\log_e (n_i + g_i - 1)\, dn_i - \log_e n_i - dn_i] = 0$$

$$\therefore \quad -d(\log_e P) = \sum_i [-\log_e (n_i + g_i - 1)\, dn_i + \log_e n_i dn_i] = 0$$

or
$$\sum_i [-\log_e (n_i + g_i - 1) + \log_e n_i] dn_i = 0 \quad ...(v)$$

As the total number of particles and total energy are constants, we have,

$$\sum_i dn_i = 0 \quad ...(vi)$$

$$\sum_i E_i dn_i = 0 \quad ...(vii)$$

Multiplying (vi) by α and equation (vii) by β and adding to equation (v), we get

$$\sum_i [-\log_e (n_i + g_i - 1) + \log_e n_i + \alpha + \beta E_i] dn_i = 0$$

$$\therefore \quad -\log_e (n_i + g_i - 1) + \log_e n_i + \alpha + \beta E_i = 0$$

Taking $n_i + g_i$ very large as compared to 1, the quantity 1 can be neglected.

$$\therefore \quad -\log_e (n_i + g_i) + \log_e n_i + \alpha + \beta E_i = 0$$

$$\log_e \left(\frac{n_i}{ng_i + g_i} \right) = -\alpha - \beta E_i$$

$$\frac{n_i}{n_i + g_i} = e^{-\alpha - \beta Ei}$$

$$\frac{n_i + g_i}{n_i} = e^{\alpha + \beta Ei}$$

$$1+\frac{g_i}{n_i} = e^{\alpha + \beta Ei}$$

$$\frac{g_i}{n_i} = (e^{\alpha + \beta Ei}) - 1$$

$$n_i = \left(\frac{g_i}{e^{\alpha+\beta E_i}}\right) - 1 \qquad \text{...(viii)}$$

Equation (ix) represents the Bose-Einstein distribution law.

Taking $\beta = \dfrac{1}{kT}$

$$\therefore \qquad n_i = \frac{g_i}{[e^{(\alpha+E_i/kT)}]-1}$$

The value of the constant a is governed by the equation $\Sigma n_i = N$. As n_i cannot be negative, α must always have a positive value. The distribution of the particles for different energy levels

SOME DEFINITIONS

It would be appropriate to discuss, at some length, the physical entities involved, in understanding the statistical problems.

A Physical Quantity

Essentially, we have two kinds of physical quantities (a) the intensive quantities that do not depend on the mass of the system, *e.g.*, P and T, and (b) the extensive quantities which depend on the mass of system, *e.g.*, S, V, E, etc.

System

It is a finite region in the space where the experimental body is kept. It Is obligatory for the system to have a finite mass. The system can be a closed one (may have a real physical boundary) or may have an imaginary boundary in space. Hence it can be penetrable or Impenetrable to the transport of energy. The properties of the particles in the system define the properties of the system.

Microstate

Each macrostate can have several microstates associated with it. Microstate is that state of the system where the individuality of the particles is not ignored. Hence the microstates are affected by change in position, momentum, energy (and mutual interaction) of individual particles. These states, however, cannot be identified by any physical experiment. Following table will help in understanding the difference between macro and microstates. We take two boxes A and B and four distinguishable particles a, b, c, d, and investigate their distribution.

It is evident from part B of the table that when the individuality of the particles is ignored, there remains no difference between the number of macrostates and associated microstates. By definition, given above, such a distribution will not give statistical probability or correct probable distribution (last column). Another illustration for the above table is to distribute 4 distinct books on 2 shelves. We have a total of 16 distributions as in part A of the table but if we distribute 4 identical books on same 2 shelves we have only 5 arrangements.

From the table, we see that for the most probable distribution (2, 2) we have 6 microstates. In other words, the macro distribution is most probable when number of microstates involved is maximum. In actual physical problems, we deal with extremely large number of particles as $\approx 10m^{-3}$ whence the number of microstates can be unimaginably large.

Macrostate

It is that state of the system which can be perceived by experiments. It is an isolated state of the system (not influenced by external agencies) and can be defined by certain value of pressure P, temperature T, and volume V. A macrostate can also be defined in terms of its microstructure, *e.g.*, by the specification of number of molecules (or phase-points in each cell of the phase space). Let us consider the distribution of 8 particles in two boxes. We can have (0, 8), (1, 7), (2, 6)... (7, 1) and (8, 0) distributions. These 9 distributions form 9 macrostates. Here we may note that a macrostate does not distinguish individuality of particles. Hence, for n particles we have (n + 1) macrostates when the distribution of particles is restricted in 2 boxes. This number will increase if the number of boxes is increased.

Phase Space

In order to specify the state of the gas from the molecular point of view, we shall know the position and momentum of each of its molecules, *i.e.*, we must specify six quantities x, y, z. p_x, p_y, p_z for each of the molecules. In three-dimensional space, we can easily consider an element of volume dx dy dz. Similarly, we can plot momentum components p_x,

p_y and p_z in three mutually perpendicular directions and again consider three-dimensional volume. In other words, let us consider a particle with only a motion of translation. Then, classically we need three positional and three momentum coordinates to describe its state. Just as the space is represented by a three dimensional positional coordinate system, we can also represent momentum space by a three-dimensional momentum coordinate system.

The state of the system then can be defined by six-dimensional positional-space and momentum space (phase) system. The two spaces together, defined by 6-dimensional (3 space and 3 momentum) hypothetical coordinate systems, is called phase space. Phase-space is purely a mathematical concept and cannot be depicted by a diagram like. So, phase-space is just a hypothetical space occupied by an infinite set of such six coordinate points. With the passage of time the point P (known as phase-point) traces out a certain trajectory known as phase-path.

As the position and momentum, in translated motion, change with lime, the phase-path may be quite complicated. Such 6-dimensional (6-d) phase space is also known as the μ-space. A large number of 6-d cells, each with a volume of $d\tau = (dx\ dy\ dz)\ (dp_x\ dp_y\ dp_z) = dV.\ dp$, together form the μ-space. Let us now pass on to a many particle system. Let the system has N particles. Then, to define their state we need 3N positional and 3N momentum coordinates in a 6-d phase-space. We can consider this problem in another way:

A single point in 3-d space represents 3 coordinates; in 6-d space represents 6 coordinates; in n-d space represents n coordinates. Hence in 6N-d space a single point corresponds to 6N coordinates. If N be the number of molecules in a kilogram-mole (or a Mole), the 6N-dimensional phase-space is called r-space (or y-space). We can generalise this representation by assuming that the system has N particles each having f degrees of freedom the Γ- space will then be 2Nf dimensional, having Nf positional coordinates (q_i, i = 1 to Nf) and Nf momentum coordinates (P_i, i = 1 to Nf).

Phase-cells

The concept of phase space enables us to understand the state of micro particles -A their position and energy. When a particle changes its momentum and position with time, there will be infinite points in the phase-space which will express the state of this moving particle. Again in a finite volume of phase space, we have infinite microstates available for the particle. Hence, we can divide the whole of the phase space into finite sized small cells. The volume of the cell, however, cannot indefinitely be small.

Accessible Microstates

Those microstates which follow stipulated conditions (like number of particles is conserved, energy is conserved, system as finite extension) are called the accessible microstates.

The State of Equilibrium

It is the steady state of an isolated system where the values of P, V and T do not change with the passage of time. In such systems, no energy is given to or taken out of it. In statistical terms, equilibrium state corresponds to that macrostate which has maximum microstates. One must, however, remember that an isolated system, in equilibrium, is equally likely to be in any of its accessible microstate.

Events

When an operation is performed or is allowed to take place, the occurrence of certain complexion, configuration or arrangement of the system, at one such trial is called an event. When number of trials is increased the probability of certain event may increase.

STATISTICAL POSTULATES

Following postulates are made for applying the statistical method to physical systems:

(i) Every system has a large number of particles/molecules which are in constant random motion.

(ii) The motion of particles/molecules is described in 2Nf-dimensional phase-space.

(iii) The phase-space is divided into cells.

(iv) Postulate of equal a priori probability holds good.

(v) Probability of a particular configuration is given by the ratio of microstates corresponding to that configuration to total accessible microstates, *i.e.*,

$P_i = \Omega_i/\Sigma\Omega_i$ and the average value of any physical parameter is given by

$$< x > = \Sigma x_i P_i = \frac{\Sigma x_i \Omega_i}{\Sigma \Omega_i}$$

where Qi is the frequency of the configuration.

(vi) The equilibrium state of the system is the state of maximum probability.

(vii) The law of conservation of mass is generally valid whence

$n_1 + n_2 + n_3 + ... = \Sigma n_i, = N =$ constant.

(viii) The law of conservation of energy governs the distribution of particles among cells.

Thus, if $\varepsilon_1, \varepsilon_2, \varepsilon_3$ ... denotes energies of each particle in cell 1, 2, 3, etc., then

$E = \varepsilon_1 n_1, + \varepsilon_2 n_2 + \varepsilon_3 n_3 + ... = \Sigma \varepsilon_i n_i =$ constant.

(ix) Every point in the phase-space (cell) represents a microstate, the cell volume being h where/is the degree of freedom.

(x) The number of microstates in the energy interval e and $\varepsilon + d\varepsilon$, per unit volume is represented as $\Omega(\varepsilon)$.

(xi) A collection of accessible microstates form a statistical ensemble.

ENSEMBLE METHOD

Ensemble method is superior to the method of *time average*, in determining average value of a physical quantity. So first we recall the characteristics of an ensemble and its way of representation.

1. It is a group of large number of systems.
2. Every member of an ensemble is an exact miniature copy of the system under test.
3. Every member of an ensemble is subjected to the same experimental conditions.
4. There may be three kinds of interactions between the members of the ensemble:

   (a) Thermal-related to T,

   (b) Mechanical-related to V, and

   (c) Diffusive-related N or m (*i.e.*, chemical potential).
5. Every member of the ensemble may have a large number of microstates (also called the dynamic state) due to random motion

of its particles/molecules/And when the configuration of the member systems is maintained same in terms of generalised coordinates and momenta, the statistical method is expected to yield quite accurate results.

Ensemble Representation

Ensembles can be represented in a Γ-space by the distribution of its phase-points. Generally, the distribution of phase-points in an ensemble is continuous and can be represented by a density function ρ(p, qt).

The number of phase-points in the vicinity of (p, q) in Γ-space is given by p (p, q t) dT where d r = dp dq = volume element of Γ- space.

If ρ (p, q t} is known at time t_1, its value can easily be obtained for some other time t_2. Now, from the probability theory

$$d\rho\ (p,\ q) \propto d\Gamma$$

$$\therefore \quad d\rho\ (p,\ q) = \rho(p,\ q)d\Gamma$$

Here ρ (p, q) is known as the distribution function. The ensemble average of any function R is then given by

$$\langle R\rangle = \int R\ (p,\ q)\ \rho(p,\ q)d\Gamma.$$

Selection of Ensemble

(i) Molecules of real gases are usually formed of 2, 3.... atoms whence the density of phase points is not uniform,

(ii) The density of phase points is non-zero for every range of energy,

(iii) Total number of particles and energy is conserved. Point (i) eliminates use of uniform ensemble. Point (ii) eliminates use of microcanonical ensemble. Point (in) eliminates use of grand canonical ensemble to real gases.

But since the properties of these gases such as N, V, T, ε, H etc., are constant the canonical ensemble finds use in description of real gases for their thermodynamic properties.

STEFAN'S LAW VERIFICATION

In 1897, Lummer and Pringsheim experimentally verified Stefan's law over a wide range of temperature (100°C to 1,300°C). The apparatus consisted of a black body C. For temperatures between 200°C and 600°C, a hollow copper sphere coated inside with platinum black was used. The fused nitrates of sodium and potassium having a melting point of 219°C were used as the bath surrounding the black body. For temperature between 900°C and 1,300°C, an iron cylinder coated inside with platinum black was used as a black body and it was enclosed in a double walled gas furnace. A thermocouple T was used as a thermometer. A bolometer B was used to measure the intensity of the emitted heat radiations. S_1, S_2 and S_3 were the water-cooling shutters. Another black body A at 100°C was used to standardize the bolometer.

The double walled vessel of the black body A contained boiling water at 100°C. The bolometer B was allowed to face the opening of the black body A and the shutter S_3 raised. The deflections in the galvanometer of the bolometer at various distances were noted and it was found that the deflection was inversely proportional to the square of the distance between the bolometer and the opening of the black body A. Thus, the deflection in the galvanometer was proportional to the intensity of heat radiations.

The shutter S_3 was closed and the bolometer B was allowed to face the opening of the black body C. The shutters of S_1 and S_2 were raised. The bath surrounding the black body was maintained at a constant temperature and the maximum deflection produced in the galvanometer of the bolometer was noted. Thus at various constant temperatures of the black body, corresponding to constant deflections (in the galvanometer of the bolometer) were observed. Then the data was reduced to a common arbitrary unit in terms of the total radiations from the black body A at 100°C. Let θ be the deflection in the galvanometer, T_1 the temperature of the black body and T_2, the temperature at the entrance of the bolometer. It was found that

$$\theta \propto (T_1^4 - T_2^4)$$

But $\theta \propto R$

$$\therefore \qquad \theta \propto (T_1^4 - T_2^4)$$

This verifies Stefan's law.

Recently Coblentz has verified Stefan's law more accurately. He took an electrically heated black body whose temperature was measured by an accurate thermocouple. An absolute bolometer was used to measure the amount of heat radiations emitted by the black body. He was able to show the correctness of Stefan's law experimentally up to 1,600°C.

GIBBS ENSEMBLE

Consider a system a such that a number of similar observations can be taken on it. Further, we assume that the system undergoes exactly similar changes in its state during each observation, *i.e.*, after every set of operations, the system returns back to its initial state (*i.e.*, the system is perfectly reversible). The propeties of ensemble was put forth by Gibbs in 1902. In his view, it was a mathematical device which provided us with an opportunity to follow a probabilistic approach in defining the state of the system at a given time.

If this be the case, we can define the state of the system in terms of its statistical probability. But this is not practically feasible as:

(i) The number of molecules/particles in the system are so large that it is not possible to have the macrostate exactly the same as it was during the preceding observation,

(ii) Each macrostate has several microstates. We cannot ascertain their identity,

(iii) Each observation may take quite a long time whence for applying statistics, several observations may need several thousand years,

(iv) To manage and process a large number of data, over years together is not possible,

(v) The system may not show reversibility to different operations.

Ensemble was a satisfactory solution to it. An ensemble is a group of N_T identical systems. These N_T systems have the same configuration, energy and accessible states.

In other words, the ensemble has N_T isolated systems each of which are exact copies of the experimental system A Thus, we have a large number of systems that behave exactly in the same manner during a set of operations as the system A Then, if N_r systems out of N_T systems produce a particular result r, we have

$$P_r = N_r/N_T$$

as the probability for that result.

If we want to measure a particular property, say x, of the system

$$< x > = \frac{n_1x_1 + n_2x_2 + ... + n_rx_r +}{n_1 + n_2 + ... + n_r +}$$

or

$$< x > = \frac{n_1x_1}{N_T} + \frac{n_2x_2}{N_T} + ... \qquad \because \Sigma n_i = N_T$$

or $\quad = P_1x_1 + P_2x_2 + ...P_rx_r + ...$

i.e., $\quad = \Sigma P_rx_r$

where x_r is the property of rth group of N_r similar systems and P_r is the probability for that result.

In this unique way Gibbs replaced the method of time averages (observations made on the same system at different times to obtain average) by ensemble average as shown above. The main advantage of the ensemble method is "the observations on all systems can be taken simultaneously under similar conditions", *e.g.*, test of quality of soaps of different brands will be quicker by millions of users.

SOLVED EXAMPLES

Example 1:

The van der Waal's constant a and b for hydrogen in litre atmosphere units are 0.246 and 2.67 × 10^{-2} respectively. Calculate the inversion temperature of hydrogen.

Solution:

we know $\quad T_i = 2a/Rb$

or
$$T_i = \frac{2 \times 0.246}{0.0821 \times 0.0267}$$
$$= 224.5^{o}k = (224.5 - 273) = -48.5^{o}C.$$

Example 2:

For equation $(P \_ a/V^2)\ V = RT$, prove that (i) dP is an exact differential, (ii) P is a state function and

$$\text{(iii)} \left(\frac{\partial P}{\partial T}\right)_V \left(\frac{\partial T}{\partial V}\right)_P \left(\frac{\partial V}{\partial T}\right)_T + 1 = 0.$$

Solution:

(i) For dP to be an exact differential, it is be proved that

$$\frac{\partial^2 P}{\partial V \partial T} = \frac{\partial^2 P}{\partial T \partial V} \qquad ...(1)$$

$$(P + a/V^2)\ V = RT$$

or
$$P = \frac{RT}{V} - \frac{a}{V^2} \qquad ...(2)$$

When the above equation is differentiated with respect to V at constant temperature T, we obtain

$$\left(\frac{\partial P}{\partial V}\right)_T = -\frac{RT}{V^2} + \frac{2a}{V^3}$$

When the above equation is differentiated with respect to T at constant V, we obtain

$$\frac{\partial^2 P}{\partial T \partial V} = -\frac{RT}{V^2} \qquad ...(3)$$

When equation (2) is differentiated with respect to T at constant volume V, we obtain

$$\left(\frac{\partial P}{\partial T}\right)_V = \frac{R}{V}$$

When the above equation is further differentiated with respect to V at constant T, we obtain

$$\frac{\partial^2 P}{\partial T \partial V} = -\frac{R}{V^2} \qquad ...(4)$$

From equations (3) and (4), we get

$$\frac{\partial^2 P}{\partial T\, \partial V} = \frac{\partial^2 P}{\partial T\, \partial V} = -\frac{R}{V^2} \qquad ...(5)$$

Thus, we have proved condition, (1). This shows that dP is an exact differential.

(ii) In the said question, first of all a change occurs in volume at constant temperature followed by a change in temperature at constant volume, thereby giving the final equation (3).

Again, a change in temperature is considered at constant volume, followed by a change in volume at constant temperature, thereby giving the final equation (4). Both equations (3) and (4) are same, thereby revealing that the final change in P remains the same irrespective of the difference in intermediate changes. Therefore, P is a state function as changes in P have been found to depend upon the intial and final states of the system.

(iii) Equation (2) is as follows :

$$P = \frac{RT}{V} - \frac{a}{V^2}$$

When the above equation is differentiated, we obtain :

$$dP = \frac{R}{V}\,dT - \frac{RT}{V^2}\,dV + \frac{2a}{V^3}\,dV$$

$$= \frac{R}{V}\,dT - \left(\frac{RT}{V^2} - \frac{2a}{V^3}\right)dV \qquad ...(6)$$

If volume is constant, dV = 0 : equation (6) becomes as follows :

$$dP = \frac{R}{V}\,dT \text{ or } \left(\frac{\partial P}{\partial V}\right)_V = \frac{R}{V} \qquad ...(7)$$

If pressure is constant, dP = 0; equation (6) becomes as follow :

$$\frac{R}{V}\,dT - \left(\frac{RT}{V^2} - \frac{2a}{V^3}\right)dV = 0$$

or

$$\left(\frac{\partial T}{\partial V}\right)_P = \left(\frac{RT}{V^2} - \frac{2a}{V^3}\right)\Big/\frac{R}{V} \qquad ...(8)$$

If temperature is constant, dT = 0; equation (6) becomes as follows:

$$\left(\frac{\partial T}{\partial V}\right)_T = -\frac{1}{\left(\frac{RT}{V^2} - \frac{2a}{V^3}\right)} \qquad ...(9)$$

On multiplying equations (7), (8) and (9), we obtain

$$\left(\frac{\partial p}{\partial T}\right)_V\left(\frac{\partial T}{\partial V}\right)_P\left(\frac{\partial V}{\partial P}\right)_T = -\frac{R}{V} \times \frac{\left(\frac{RT}{V^2} - \frac{2a}{V^3}\right)}{\left(\frac{RT}{V^2} - \frac{2a}{V^3}\right)} \times \frac{1}{R} = -1$$

or

$$\left(\frac{\partial p}{\partial T}\right)_V\left(\frac{\partial T}{\partial V}\right)_P\left(\frac{\partial V}{\partial P}\right)_T + 1 = 0$$

This is cyclic rule.

Example 3:

For an ideal gas PV = RT, prove that dP is an exact differential.

Solution:

We know $PV = RT$

or

$$P = \frac{RT}{V} \text{ (R is a constant)} \qquad ...(1)$$

On differentiating with respect to T, taking V constant, we obtain

$$\left(\frac{\partial P}{\partial T}\right)_P = \frac{RT}{V} \qquad ...(2)$$

Again, on differentiating with respect to V, we obtain

$$\frac{\partial^2 P}{\partial V \partial T} = -\frac{R}{V^2} \qquad ..(3)$$

On differentiating (1) with respect to volume, taking T constant, we obtain

$$\left(\frac{\partial P}{\partial V}\right)_T = -\frac{RT}{V^2} \qquad ...(4)$$

Again, on differentiating above equation with respect to temperature taking V as constant, we obtain

$$\frac{\partial^2 P}{\partial T \partial V} = -\frac{R}{V^2} \qquad ...(5)$$

From equations (3) and (5), we obtain

$$\frac{\partial^2 P}{\partial T \partial V} = \frac{\partial^2 P}{\partial T \partial V}$$

Thus dP is an exact differential.

Example 4:

A thermostat was maintained at 370.05 K. For nearly an hour 4.2 kJ of heat leaked through the thermostat insulation into a room where the initial temperature of air was 300.05 K. (i) What was the entropy change of the material in the thermostat ? (ii) What was the entropy change of the air in the room ? (iii) Was the process spontaneous? (Material in the thermostat is water).

Solution:

(i) Entropy change of the material in the thermostat

$$-\frac{4.2\ \text{kJ}}{370.05\text{K}} = -\ 11.35 \times 10^3\ \text{kj K}^{-1}$$

(ii) Entropy change of the air in the room

$$= \frac{4.2\ \text{kJ}}{300.05\ \text{K}} = 14.00 \times 10^3\ \text{kJ K}^{-1}$$

(iii) Now, $\Delta S_{total} = \Delta S_{thermostat} + \Delta S_{room}$

$$= (-\ 11.35 + 14.00)\ \text{JK}^{-1}$$

$$= 2.65\ \text{J K}^{-1}$$

Since ΔStotal is positive, the process of leaking is spontaneous.

Example 5:

Calculate ΔS, ΔH and ΔU for the process

$$H_2\,O(l,\ 20^oC,\ 1\ atm \rightarrow H_2O\ (g,\ 250^oC,\ 1\ atm)$$

Given the following data :

$$C_P\ (l) = 75.6\ J\ K^{-1}\ mol^{-1}$$

$$C_P\ (g) = 36.2\ J\ K^{-1}\ mol^{-1}$$

ΔH for vaporization of H_2O at 100°C and 1 atm is equal to 40.85 k J mol^{-1}.

Solution:

The given process may be replaced by the following processes.

(i) H_2O (l, 20°C, 1 atm) → H_2O (l, 100°C, 1 atm)

(ii) H_2O (l, 100°C, 1 atm) → H_2O (g, 100°C, 1 atm)

(iii) H_2O (g, 100°C, 1 atm) → H_2O (g, 250°C, 1 atm)

Thus we have

$$\Delta S = \Delta S_{(1)} + \Delta S_{(ii)}\ \Delta S_{(iii)}$$

$$= C_{p,m}(1)\ \ln\frac{T_2}{T_1} + \frac{\Delta H_{vap}}{T} + C_{p,m}(g)\ \ln\frac{T_2'}{T_1'}$$

$$= \left(75.6\ \ln\frac{373}{293} + \frac{40850}{373} + 36.2\ \ln\frac{523}{373}\right) JK^{-1}\ mol^{-1}$$

$$= (18.25 + 109.52 + 12.24)\ J\ K^{-1}\ mol^{-1}$$

$$= 140.01\ J\ K^{-1}\ mol^{-1}$$

$$\Delta H = \Delta H_{(1)} + \Delta H_{(ii)} + \Delta H_{(iii)}$$

$$= C_{p,m}\ (l)\ (T_2 - T_1) + \Delta H_{vap} + C_{pm}(g)\ (T'_2 - T'_1)$$

$$= (75.6 \times 80 + 40850 + 36.2 \times 150)\ J\ mol^{-1}$$

$$= 52328\ \ J\ mol^{-1}$$

$$\Delta U = \Delta U_{(1)} + \Delta U_{(ii)}\ \Delta U_{(iii)}$$

$$= C_{v,m}\ (l)\ (T_2 - T_1) + \Delta U_{vap} + C_{v,m}(g)\ (T'_2 - T'_1)$$

$$= C_{p,m}(l)(T_2-T_1)+\{\Delta H_{vap}+\Delta_{V\ g}RT\}+\{C_{p,m}(g) - R\}(T'_2 - T'_1)$$

$$dw = nR\ dT - V\ dP = nR\ dT - \frac{nRT}{P} dP$$

Now $$\frac{\partial}{\partial P}(nR)_T = 0$$

and $$\frac{\partial}{\partial T}\left(-\frac{nRT}{P}\right)_P = -\frac{nR}{P}$$

Hence, dw is inexact differential.

Suppose f = 1/T be an integrating factor. Then

$$f\ .\ dw = \frac{1}{T} nR\,dT - \frac{nR}{P} dP$$

$\therefore$ $$\frac{\partial}{\partial P}\left[\frac{nR}{T}\right]_T = 0$$

and $$\frac{\partial}{\partial T}\left[\frac{nR}{P}\right]_T = 0$$

Thus f. dw $\frac{nR}{T}$ dT $-$ $\frac{nR}{P}$ dP is an exact differential while 1/T is an integrating factor.

Example 9:

Derive an expression for the entropy change when a van der Waals gas is heated and expanded simultaneously.

Solution:

Taking S = f(T, V), we get

$$dF = \left(\frac{\partial S}{\partial T}\right)_V dT + \left(\frac{\partial S}{\partial V}\right)_T dV$$

$$= \frac{nC_{V,m}}{T} dT + \left(\frac{\partial P}{\partial T}\right)_V dV$$

Now for van der Waal's gas we have

$$P = \frac{n\ RT}{V - nb} - \frac{n^2 a}{V^2}$$

Therefore $$\left(\frac{\partial P}{\partial T}\right)_V = \frac{n\ RT}{V - nb}$$

Thus $$dS = \frac{n\,C_{V,m}}{T}\,dT + \frac{n\,R}{V-nb}\,dV$$

Which on integrating within the limit T_1, V_1 and T_2, V_2 gives

$$\Delta S\ nC_{V,m} \ln \frac{T_2}{T_1} + nR \ln \frac{V_2 - nb}{V_1 - nb}$$

Example 10:

For an adiabatic process, dq = 0, If one were to write

$$\Delta S = \int \frac{dq}{T} = \int \frac{0}{T} = 0$$

Then every adiabatic process would be iso-entropic process. Comment.

Solution:

For an adiabatic process, dq – 0, and according to the first law of thermodynamics, dq = dU – dw

Thus, we will have

$$dU = dw$$

The expression of entropy change will be given as

$$\Delta S = \int \frac{dU - dw}{T}$$

$$= \int \frac{dU}{T} + \int \frac{P_{opp}}{T}\,dV$$

Now, it is only for reversible process the first integral is equal and opposite sign of the second integral and thus $\Delta S = 0$. For an irreversible process, these two integrals do not have equal magnitude and thus sum is not zero. Hence $\Delta S \neq 0$ for an adiabatic irreversible process.

Example 11:

Calculate the maximum work obtainable by (i) the isothermal expansion and (ii) the adiabatic expansion of 2 mol of on ideal gas initially at 25° C form 10 L to 20 L. Assume Cv = (5/2) R.

Solution:

(i) Maximum work is obtained in reversible expansion. Hence

$$w = -\ nRT \ln \frac{V_2}{V_1}$$

$$= -\ (2\ mol)\ (8.314\ J\ K^{-1}\ mol^{-1})\ (298\ K)\ 2.303\ \log \frac{20L}{10L}$$

$$= -\ 3435\ J$$

(ii) Firstly, we calculate the final temperature using the expression

$$T_2\ V_2^{\ \gamma-1} = T_1\ V_1^{\ \gamma-1}$$

where $$\gamma = \frac{C_P}{C_V} = \frac{(7/2)R}{(5/2)R} = \frac{7}{5}$$

Hence, $$T_2 = (298K)\left(\frac{10L}{20L}\right)^{(7/5)-1}$$

$$= (298\ K)\ (0.7578)$$

$$= 225.8\ K$$

Now $$W_{adi} = \Delta E$$

$$= n\ C_{V},\ m(T_2 - T_1)$$

$$= (2mol)(2.5 \times 8.314\ JK^{-1}mol^{-1})(225.8K - 298K)$$

$$= -\ 3001\ J.$$

Example 12:

How much heat is required to raise the temperature of 1 mole oxygen from 300K to 1300 K at constant pressure.

$$C_p = 6.095 + 3.325 \times 10^{-3}\ T - 1.017 \times 10^{-6}\ T^2.$$

Solution:

$$H_2 - H_1 \int_{T_t}^{T_2} C_p\ dT$$

$$T_1 = 300\ k,\ T_2 = 100\ k$$

$$\therefore\ H_{1300} - H_{300}$$

$$= \int_{300}^{1300} \left(6.095 + 3.253 + 10^{-3}\ T - 1.017 \times 10^{-6}\ T^2\right)$$

[Substituting the value of C_p]

$$\because\ H_{1300} - H_{300} = 6.095\ (1300 - 300) + \frac{3.253 \times 10^{-3}}{2}$$

$$\times\ [(1300)^2 - (300)^2] - 1/3 \times 1.017 \times 10^{-6}\,[(1000)^2 - (300)^2]$$

$$= 6095 + 2602.4 - 735.63$$

$$= 7961.77 \text{ cal/mole.}$$

Example 13:

Show that

$$\left(\frac{\partial \alpha}{\partial P}\right)_T + \left(\frac{\partial \beta}{\partial T}\right)_P = 0$$

where $\propto$ and β are coefficient of thermal expansion and coefficient of compressibility, respectively.

Solution:

Since V is a state function, we can write

$$\frac{\partial^2 V}{\partial T\,\partial P} = \frac{\partial^2 V}{\partial P\,\partial T}$$

or $$\frac{\partial}{\partial T}(-V\beta)_P = \frac{\partial}{\partial T}(V\alpha)_T$$

Carrying out the differentiation, we get

$$-\beta\left(\frac{\partial V}{\partial T}\right)_P - V\left(\frac{\partial \beta}{\partial T}\right)_P = \alpha\left(\frac{\partial V}{\partial P}\right)_T + V\left(\frac{\partial \alpha}{\partial P}\right)_T$$

that is $$\left(\frac{\partial \alpha}{\partial P}\right)_T + \left(\frac{\partial \beta}{\partial T}\right)_P = -\frac{\beta}{V}\left(\frac{\partial V}{\partial T}\right)_P - \frac{\alpha}{V}\left(\frac{\partial V}{\partial P}\right)_T$$

$$= -\frac{\beta}{V}(\alpha V) - \frac{\alpha}{V}(-\beta V) = 0$$